8/04
30.95

D1237995

The Craggy Hole in My Heart and the Cat Who Fixed It

This Large Print Book carries the
Seal of Approval of N.A.V.H.

The Craggy Hole in My Heart and the Cat Who Fixed It

Over the edge and back with my dad, my cat, and me

Geneen Roth

WHEELER
PUBLISHING

Published in 2004 by arrangement with Harmony Books, a division of Crown Publishers, Inc.

Wheeler Large Print Compass.

The text of this Large Print edition is unabridged. Other aspects of the book may vary from the original edition.

Set in 16 pt. Plantin by Al Chase.

Printed in the United States on permanent paper.

ISBN 1-58724-771-2 (lg. print : hc : alk. paper)

To my mother,
Ruth Wiggs
For love

As the Founder/CEO of NAVH, the only national health agency solely devoted to those who, although not totally blind, have an eye disease which could lead to serious visual impairment, I am pleased to recognize Thorndike Press★ as one of the leading publishers in the large print field.

Founded in 1954 in San Francisco to prepare large print textbooks for partially seeing children, NAVH became the pioneer and standard setting agency in the preparation of large type.

Today, those publishers who meet our standards carry the prestigious "Seal of Approval" indicating high quality large print. We are delighted that Thorndike Press is one of the publishers whose titles meet these standards. We are also pleased to recognize the significant contribution Thorndike Press is making in this important and growing field.

Lorraine H. Marchi, L.H.D.
Founder/CEO
NAVH

★ Thorndike Press encompasses the following imprints: Thorndike, Wheeler, Walker and Large Print Press.

There is a crack in everything,
that's how the light gets in.
— LEONARD COHEN

One

When my friend Sally called to tell me that I needed a kitten, and fortunately, her cat Pumpkin was pregnant, I said no, absolutely not. I didn't want a pet, I didn't like cats, and I didn't want to love anything that could die before me.

I was thirty-three years old, single, and living alone in a house with a garden, three leaky skylights, and a crooked path to a sheltered beach in Santa Cruz, California. After seventeen years of struggling madly with emotional eating, and being as insane as anyone I'd ever met — I'd gained and lost over a thousand pounds — I'd finally crawled out of the compulsion by giving up dieting altogether. More recently, I'd settled at my natural weight, written two books, and begun teaching national workshops about breaking free from emotional eating.

But my obsession with food was a walk in the park compared to the chaos that ensued whenever the possibility of love walked into my life. At the time of Sally's call, I was in a "relationship" — I use that term loosely — with Harry-the-Rake, a self-confessed lothario, who alternated between wanting to move in with me and telling me I was too fat. I was convinced that my heart was either on permanent sabbatical or missing some essential ingredients — the ones that allowed normal people to take risks, to discern the bad guys from the good, to say come closer, hold me, go away. And I was wary of opening to anyone or anything that would depend on me to come through. I didn't trust myself to show up. I didn't think I had the capacity for big love.

Pumpkin gave birth to two kittens whom Sally immediately named Blanche and June. My mother, visiting from New York at the time, wanted to see them. At two hours old, they looked like wet weasels, and I wasn't impressed. My mother went straight for the white kitten. Take this one, she crooned, as she stroked the slicked-back fur of the shut-eyed rodent, but I wasn't taking anything so fast.

A few weeks later, Sally called and said her husband didn't want a white cat, and so Blanche was mine. Usually, I am the one who bosses people around, but Sally was completely sure of herself, absolutely positive that having this pet was a precursor to having a life. So I told her I would take the kitten on one condition: if I didn't like being a cat mother, I could return it in two weeks, like a pair of gloves from Macy's. She agreed.

10

It's not that I'd never had a pet. My grandmother gave me a parakeet named Cookie when I was seven. She rode around the house on my shoulder, sat on the desk while I did homework, and pecked at my eyelashes when I closed my eyes. One day, my brother opened the front door and Cookie flew out of the house. I cried for weeks. I decided then that the next thing I loved was not going to be able to fly away. We settled on goldfish, but the one we called Tallulah got out of the bowl somehow and flipped around the house. My mother and I ran after her with a strainer, but we couldn't catch her, and she died under the brown paisley couch. Then there was a puppy named Cocoa, who pooped in my father's slipper right before he stepped into it one Sunday morning, and by Monday, she had gone to live somewhere else.

When she heard that Sally wanted to give me a kitten, my friend Sophie told me her pet story. After her mother died and her husband left her for another woman, she thought she was going crazy — the kind of crazy where a psychotic break was two weeks away. On a particularly rough day, a group of friends tried to make her feel better, but she sensed their fear. The fact that her best friends couldn't be with her sorrow made her feel even more frightened, more alone. Then her dog, Squeak, jumped in her lap and fell

11

asleep. In that moment, she says Squeak saved her life. He cut through the drama, walked directly on the fiber of feelings, and stayed there, as if pain and grief were no big deal — as natural as chasing squirrels. His relaxation dissolved her fears of going crazy. After that, she was left with a broken heart, and as much as that hurt, she knew it would mend.

Though I was glad Sophie had her dog, I'd heard these sappy tales before — a boy and his dog, a girl and her parrot, the wolf who saved the family from a fire — and didn't see what they had to do with me. I still didn't want a cat.

During our first few days together I refuse to be charmed by Blanche, although every time I turn a corner, she is there, crouching behind philodendron leaves, or stalking an ant or a dust mote or my big toe. When I say no, she doesn't hold a grudge. When I push her away, she comes back. Blanche's affection doesn't waver if my hair sticks straight up in the mornings or if I am having a fat day. She seems to be looking beneath the surface of things at some backward-spreading light I am not aware of.

A week after Blanche arrives, my two-year relationship with Harry-the-Rake ends when he falls in love with another woman. Flinging myself on the bed in a paroxysm of sorrow

— what will I do, where will I go, who will ever want me — I notice a cloud of fluff inching across the quilt until it settles on my heaving chest, wheezing a low, gravelly purr. It's difficult not to be melted by such total acceptance; it's hard to keep insisting that the world is a terrible place.

On the eleventh day, I admit I am smitten and tell Sally I will keep the cat.

Once I cross over, every single thing about Blanche enchants me, and I am positive that no one has ever had a cat this adorable. Then I start to worry that I love her because all kittens are irresistible, but when she gets older, I won't love her anymore. I still believe love depends on what you look like.

Within a month, Blanche has about ten thousand nicknames: Pooters, Banana, Wig-Wig, Moochy-Mooch, Fuzzy-Wuzz, Petunia, Mr. Guy and a Half, Sweet Potato, Booch Pie, Blue, Moo, Dandelion, Blanchebananche, Peachy Canoe and Tyler Too, Curly-Whirl, and on and on. Every day, a different name.

Within two months, I can't imagine that I've ever lived without her. She seems to be exactly the same shape as the craggy hole in my heart, so when I see her, all my stick-out edges and weird crazy ways smooth down. I feel as if I've been dreaming her for years and now she is here.

It never occurs to me to question my

choice of love objects or wish that Blanche was a person instead of a cat. When you've been famished for decades and someone hands you a slice of warm pumpernickel raisin bread and homemade jam, you don't ask for chocolate cake instead.

The first time she visits the vet, we discover that Blanche is a he. Since I have been calling him her, and since he has a girl's name, it is perplexing to discover the truth about Blanche's gender. But there is no question about changing his name; the being in this cat's body is definitely a "Blanche."

Dr. Mike reminds me of the popular sixties song by Johnny Cash called "A Boy Named Sue." I decide that since Blanche is going to be neutered, he has transcended gender. He is neither cat nor person, neither boy nor girl. Blanche is beyond definition.

When friends walk into my house and see that I have a kitten, they turn to mush immediately, talk baby talk, tell Blanche they love him. My friend Nancy, a suit-clad district attorney, crawls around on her knees, trying to lure him with a penguin stuffed with catnip. My painfully shy friend Louis pulls a string on the floor, from room to room, letting Blanche pounce on it. My hip, edgy friend Maria picks him up, cuddles him, and coos, ignoring me altogether. People change around him, the way they do around babies. Blanche seems to provide an opening from which their love, coiled like a rope at the bottom of a basket, can wave its vulnerable, tender head.

By the time he is two years old Blanche weighs twenty pounds. He looks like a furry pyramid or a goat with curly stomach hair. Since my books are about emotional eating, everyone who walks in the house has a comment about his size. They all say the same things:

Your cat needs to read one of your books.

Your cat needs to come to your workshops.

Your cat needs to go on a diet, but oh yeah, I forgot, you don't believe in dieting.

It doesn't help that Blanche has a girl's name and I have to keep correcting everyone that she is a he. They take it as an opportunity for further speculation: Does he eat because he's confused about his identity?

But I know this is Blanche's real shape, his natural weight, since I only feed him half a cup of dry food a day, plus little bits of butternut squash, sweet potatoes, and dried sardines. Blanche is a nibbler, a delicate eater, an epicure.

He is also the kind of cat you can dress up in a bonnet and wheel around in a baby carriage, which my eleven-year-old neighbor, Rosie, does several times a week. As soon as you pick him up, he relaxes his body and purrs; when Rosie isn't out wheeling him up and down the block, I walk with Blanche around my neck like a monkey, like a second heart.

I feel like a cliché. For the first time in my life, I am not afraid of being too intense, too effusive, too needy. No matter how many times I kiss him, hug him, pull his tail, and turn him upside down, he doesn't turn away. Blanche is a love sponge with a thousand petal-pink lipstick marks on his head.

Three months after Blanche's second birthday, I meet Matt at the Association for Humanistic Psychology conference, where we are both speakers. Though he is sexy, funny, kind — and here's the linchpin: AVAILABLE — he needs to pass the Blanche test before I let him into my life.

When Matt comes to my house on our first date, Blanche is out carousing in the neighborhood. Matt and I sit in the blue striped chairs on the deck and tell each other about our lives, the usual first-time stories. We discover that we had been to movies at the same theater in Fresh Meadows, New York, and must have passed each other on the lines for *Dr. Zhivago* and *A Hard Day's Night* when we were in high school. I tell him I didn't think I would have liked him, though — he is too nice, and I only liked boys who were mean and loved someone else. He happens to mention that he doesn't like chocolate, and I wonder whether I can ever love him.

A few seconds later, Blanche comes hopping over the fence, swaggers to Matt, and jumps on his lap. I am sorry I haven't asked Matt if he has a hernia, because when Blanche lands on you, it feels as if a truck has crashed on your legs. Matt doesn't flinch. He begins to talk baby talk.

Then, looking at me, he says, "You know, I really don't like cats."

I glance at my watch to see when I can kick him out.

"But there is something very unusual about you, Blanche," he continues, stroking him under the chin. "You seem to be more than a cat."

I decide to wait a few weeks before I ask him to marry me.

After our first date, Matt flies off to Hawaii on a business trip, and I get ready to go to New York to teach. As a treat for Blanche, and because I feel guilty about leaving him the next day, I open a can of tuna fish, and when he doesn't come tearing to my side, I know that something is wrong. I call the vet to tell him that Blanche is dragging his bottom across the deck and won't eat his favorite food. Dr. Mike tells me to bring him in immediately; he says it sounds as if Blanche has a blocked kidney.

Fortunately, my assistant, Maureen, who is working in the house, has a three-year-old child and is practiced at being calm in emergencies, because I am suddenly hysterical and can't remember where I put the cat carrier. We end up wrapping Blanche in a towel, tearing out of the house, honking through red lights, and running into the vet's office. Dr. Mike feels Blanche's kidneys, asks me when he peed last (I have no idea), and confirms the diagnosis: feline urinary disorder, a condition common in male cats.

"A few more hours and you would have lost him," he says, "his kidneys would have burst." Since by now I cannot imagine life without Blanche, I put all my emotional energy into setting up a visiting schedule for Blanche's upcoming week in the hospital. Each day a different friend will read or sing to him, bring a stuffed toy or catnip, and call me in New York so Blanche can hear my voice. It is the calling-me-in-New-York part that makes it apparent I've gone over the top.

Back at home, my feelings for Matt grow stronger, which is becoming a problem. Not only am I, a self-proclaimed curmudgeon, unexpectedly and boundlessly attached to a cat who is probably going to die before me, I am now falling for a human as well, and it scares me. I worry I'll get soft around the edges, begin getting used to his smell, the lilt of his voice, the crinkles around his eyes — and then wham! I could lose him. He could meet someone else (someone nicer, someone less intense, someone with big hair and long legs) on the street, in an airplane, at the grocery store, and break my heart. Or he could die in a plane crash, or a car accident, or from cancer. The statistical odds are against us. Men die before women. I feel utterly exposed, as if I am peeling back my skin and opening myself to the center where wounds are born.

Avoiding this state is the very reason I was obsessed with food for seventeen years, the reason I used to zing up and down the scales by ten pounds every few weeks. It seemed to me that being thin was like wearing my insides on my outside, while being fat gave me protection. People thought they were seeing me but I knew they were seeing my fat; I was safely inside, watching, waiting, assessing the situation. When they rejected me, they were only rejecting my fat. The real truth was, they couldn't touch me, which was exactly what I *wanted*. I was able to stop eating compulsively, in part, by telling myself that being thin didn't have to mean relinquishing my control about who touched me, who hurt me, who came close, and who stayed away.

It worked.

I lost weight, and until now have not been with anyone who could truly hurt me. My relationships with Harry, and Michael before him, don't count. Loving someone who is emotionally unavailable is the same thing as using fat to barricade yourself. There's no real chance at intimacy, no real vulnerability. What is closest to the bone, what is rawest and deepest and truest, never gets seen or touched.

Why love someone who is just going to turn around and either leave or die? I don't get it.

What do other people know that I don't? How can they open themselves, petal by petal, until they are completely revealed? Don't parents live with their hearts in their mouths every time their kids walk out the door? Isn't it better, since sorrow is inevitable, not to invite it in?

My nights are already sleepless knowing Blanche is roaming the neighborhood; I am certain he is going to be catnapped or hurt in a fight or suddenly decide that living somewhere else is better. I cannot tolerate the thought of losing him; why should I double my potential grief by letting Matt into my life?

However, something (grace? insanity?) seems to be pulling at me to melt, merge, toss away years of resistance. My defenses feel flimsy and insubstantial, like a papier-mâché wall trying to stop a hurricane, and though I tell myself I do have a choice — I can tell Matt I am better off alone — the thought of living without love when I can have it is worse than imagining a future when I've lost it.

According to psychological literature, we are all born with unconditional trust in the benevolence of the universe; we know we belong here, that the world is good, and that even when things are painful, they work out for the best. If, as an infant, you lived in an environment where you were welcomed, your experience of basic trust was not disrupted: you were fed when you were hungry, changed when you were wet, and cuddled when you needed to be held.

But if your mother was sad or sick or gone, and your father was working or unavailable or away, frustration and fear became your constant companions. You learned that being loved (and staying alive) required manipulating other people or controlling the situation. You stopped believing that if you relaxed, things would turn out okay, because they rarely did.

In my work with emotional eaters, lack of basic trust shows up in what I call "the one-wrong-move syndrome." It's the feeling that when you take one bite of chocolate, you might as well eat the entire box, the ice cream in the refrigerator, the twelve boxes of cookies in the cabinets, and keep on going until you've finished everything in the house,

21

because it was always just a matter of time until chaos took over, and now that it has, you can finally stop pretending that it can ever be any different. Lack of basic trust is what causes usually rational people to want to throw books at me when I mention that if they trusted their hunger it would not betray them, because they learned very early that their needs were not acceptable.

Though the one-wrong-move syndrome no longer dictates what I eat, it still controls how I love.

Despite my ambivalence (which shows up during fights when I tell Matt it's fine with me if we split up or when I think it's too messy to be close to someone so why bother), Matt's buoyancy is constant. Being with him is like basking in living daylight; since we met, my lows are higher and my highs are steadier.

Matt moves in and takes up parenting with gusto. Whenever Blanche waddles into the room, our conversation stops and we watch him move as if we have never seen him before and are utterly dazzled by the presence of so much splendor in one location. Since Matt still maintains that he doesn't like cats, he believes the way Blanche looks, walks, the way he stops halfway toward you, sits down, blinks, thinks about a few hundred things, then keeps walking is utterly original. Also,

Matt gets quite offended when anyone calls Blanche big or fat. He counters with "Big? Compared to what? A horse? A giraffe? Blanche is TINY compared to most animals."

Like any proud parent, Matt dismisses Blanche's foibles. Every night, Blanche jumps from the shelves in our bedroom to the higher shelves in the bedroom to the open-air attic above the shelves, where he hunkers down among the boxes. In the middle of the night, he needs a change of scenery, but instead of following his ledge-climbing routine in reverse, he takes a flying leap to the middle of our bed, twelve feet down.

The first time this occurs, Matt and I think The Big One — the earthquake northern California has been expecting since 1906 — has finally erupted. We tear out of bed and begin frantically looking for Blanche so that the three of us can huddle under the desk or beneath the door frame while the rest of our house collapses. After our hearts stop banging, we realize nothing except the bed has moved. Then we see Blanche in the middle of the quilt — flat on his back with all four paws up in the air — exhausted after his workout on the feline trampoline. During the next few weeks, we try piling books on the shelves so that he can't ricochet to the top, but he manages to sabotage our efforts and crash onto the bed like clockwork at two a.m.

I am furious. I threaten to refuse to give

Blanche dried sardines, but Matt is charmed. His work in the world is to teach company executives how to laugh and play, and he is convinced that these nocturnal launches prove our cat has a sense of humor.

Those of us who don't have basic trust can learn it, but it takes time and patience. First, you have to be aware that the way you see the world isn't necessarily the way it is, and second, you have to believe that you have a choice about which reality you perceive.

Once, when Blanche was infested with fleas, Matt and I listened as the vet told us how we needed to fumigate the house: remove all the plants; cover the computers, stereos, telephones, and televisions; vacuum all the carpets; wash all the sheets and towels; put flea bombs in the house; and then run out the front door for a few hours and let the poison do its work.

As we were walking out to the car, Matt and I looked at each other. Simultaneously, he grinned and said, "No big deal," as I groaned and muttered, "What a nightmare."

I was suddenly infuriated by his optimism, disgusted by his sunny-ever-after outlook.

"What planet are you living on, anyway? You live in constant denial about how difficult things are. I am FED UP with carrying the dark side for the both of us. Wake up! Life is hard."

Matt started to laugh, which infuriated me more. "It's hard, honey, but not as hard as you make it. Blanche has fleas, so what? We vacuum, we wash, we drop the flea bombs, we run out the front door. Nothing terrible is happening. It's your reaction to what's happening that makes it terrible."

"I HATE it when people say that," I said. "As if I have a choice about my reactions. *Please.* No more New Age psychobabble."

It is impossible to see what Matt sees — golden fields and lush mountain passes and wildflowers everywhere; admitting that beauty and goodness are true and real means facing everything that keeps me from seeing them. It means going back into those old, frozen feelings that obscure my vision. I don't want to go there. Not again. I've already spent years in therapy and have come far enough to allow at least some measure of love in my life. And if my resistance to pain isn't enough, there is the fact that Matt is a person, not a cat, which makes it impossible to stop comparing myself to him, to the fact that his mother adored him, made him shoes, and breast-fed him at a time when the doctors thought it was bad for her health.

Blanche is my oasis. I don't feel crazy next to him. I don't feel as if I need to measure up to his standards. I don't feel envious of his family of origin (even at my most neurotic, I wouldn't want Pumpkin for a mother). I can marvel at the way he finds the sunniest place in the house, stretches out his paws like he is surfing on light, gives one long contented sigh, then falls asleep — all of it without apology, as if he unquestionably deserves love, warmth, beauty, affection, quiet, rest, contentment. As I watch him, which I do continually throughout the day, it occurs to me that it is possible to live in the world without the low-level anxiety, frustration, and hypervigilance I've come to associate with being alive. I resist allowing Matt to be a model of trust and optimism, because he is too much the same as me. But animals are different. They teach by subversion, by making it look as if they are just pets.

When I worked with a survivor of child sexual abuse, she told me that after her father raped her nightly, she'd hide in the barn for safety. She said the mice became her friends; she realized they could bite her, but they were the only softness she knew, the only beings who might not betray her. Every night, she held them and whispered to them, especially the two whom she named Minnie and Daisy. It was with mice, she said, that she learned about love.

After we've been together for four years, Matt asks me to marry him. I say yes, even though three minutes later, I find myself thinking I am probably a lesbian and our

26

whole relationship is a terrible mistake. But I soon realize that this is my "I am someone for whom love never works and so if this is working, it has to be wrong" theme, and somehow, that makes it easier to dismiss.

Nevertheless, it is becoming obvious that it is difficult for me to accept Matt's love. It's as if I am wearing a rubber overcoat and though he showers me with sweetness, it never touches my bones. I make the motions, I say the words, I know enough not to ruin this by running off with another man, but I can't seem to let myself be deeply loved or love him without reservation.

Years later, my teacher Jeanne will tell me that I have been spending my life protecting myself from losses that have already happened, but it doesn't feel that way. It feels as if loving is dangerous in every moment. As if letting down my guard is like agreeing to be destroyed.

Except with Blanche. Every folded-down corner of love I have never let myself feel I feel with him. Blanche is like food once was — he doesn't talk back, he doesn't hit, he doesn't go away, he doesn't abuse. Also, and this is important, he doesn't have any calories.

When Matt and I get married, we place a picture of Blanche at the very front of the temple. At one point during the ceremony, I notice his picture has fallen down and ask the rabbi to stop speaking until we can set it right. This causes quite a stir in the crowd. Now everyone — his family, college friends, my longtime friends — worries that Matt is marrying a crazy cat woman who will grow old with chin hairs and swollen ankles and dozens of cats pouring themselves like warm syrup on her balding head. My best guess is that he probably is.

The three of us — Matt, Blanche, and I — move to Berkeley, where we find a veterinarian who is an acupuncturist, a homeopath, a surfer, and a saxophone player. Dr. Cheryl Schwartz talks to Blanche — whom she calls Mister Blanche — as if he can understand every word, because, she says, he does. She tells him that his kidneys are weak, and that his excessive weight puts pressure on his whole system. When she listens to his chest with a stethoscope, he has so much fur and flesh she can't find his heartbeat.

For fleas, Dr. Cheryl Schwartz recommends green tea and aloe vera gel; for bruises, a homeopathic dose of arnica; for indigestion, two little black balls of Chinese medicine we have to find in Oakland Chinatown. She tells me that Blanche is my "familiar," and that he is more connected to me than any being, ever. She says when I get upset, he gets upset. When I am sad, he is sad. She says that animals take on diseases for people, that after their people were diagnosed with cancer or other illnesses, perfectly healthy cats and dogs would become sick and the people get better. She tells me that Blanche takes care of me; he sees it as his job. And when the inevitable subject of my work and his weight arises, she concurs that it is no accident that he is one of the fattest cats she's ever met.

Blanche hates her; he glowers and glares and complains, but we follow her instructions about everything, halve his food intake, and take him for regular acupuncture treatments every six weeks, though her office is forty-five minutes away from our house.

Nothing is too much, too excessive, too time-consuming where Blanche is concerned. The tight-fisted place in my chest is relaxing, and although I know it isn't just because of my cat, I am beginning to believe I belong in any world with Blanche in it. The caveat is that he needs to *stay* in it, and he is now a ten-year-old cat with weak kidneys and a weight problem.

No matter how many precautions we take, I begin to wake up a few times a week in a cold sweat dreaming that Blanche is lost or dead. If he is not sleeping on the bed, I roam the house until I find him curled on top of the heater or in the laundry basket and drag him back to bed. When Matt and I go to Hawaii for a vacation, I dream that Blanche is dying and I am so certain it is a premonition, not a fear, that I call the house-sitter once a day to check on him. Matt is patient with my anxiety-cum-obsession, but we are both aware that something in my psyche is terribly amiss.

Something is — and it doesn't have to do with Blanche.

I know this territory. I've lived here most of my life. Blanche and Matt are passports that allow me to leave for extended vacations, but this is the place I call home. My Blanche dreams are almost exact replicas of the dreams I've had about my father for thirty years.

Two

My father.

Everyone said we looked exactly alike because our faces were shaped like moons and our lips like hearts. When I had bad dreams as a child, I would pad to his side of the bed, and tug on his pale blue pajama sleeve. He'd wake instantly, fold me in his arms, and tell me silly stories until I fell back asleep.

My mother threw things and cried, but my dad was soft as a cloud, and I twirled myself like a vine around his silvery love.

At six years old, I began having nightmares that he would die. Once he got run over by a train. Sometimes he had a heart attack or drowned. My dreams reached a fevered pitch in my twenties and early thirties; as soon as it got late enough to reach him at home in New York, I dialed his number. "I'm fine, sweetheart," he'd say. "You should be in bed. It's three a.m. there in California."

My father had hundreds of names for me, more than I have for Blanche. Mrs. Gabeanie, Wendell, Charlie Jackson, Ollie, Mabel, Gov'ner. But his favorite was Pussycat. He sung it in a high voice and rolled

31

his eyes at the same time, twirled his arms, and began shuffling his feet, as if he was about to launch into a soft shoe. All this just to greet me.

When I was twelve, he came home from work one night with a black leather-bound ledger. Inside were thick vellum pages of green-and-white stock certificates, printed with the name "The Pussycat Club" in huge black letters. Scrolls and flourishes, eagles and loopy writing took up every inch of the official documents. "Omigod," I squealed, "real live stock certificates!" My father explained that I was now the president of a corporation, with the awesome responsibility of deciding upon the stockholders. The only qualifications were: number one — I had to like them; number two — they had to be nice to me; and number three — they had to kiss him on the cheek. For the next month or so, the Pussycat Club certificates were the hottest item in town, the girly equivalent of Mickey Mantle baseball cards. And man, I was stingy with the shares. To Barbara, the baby-sitter, I issued four shares, to my friend Geri, who maintained she knew my mother longer than me because her birthday came before mine, I gave half a share. It was like being made princess, and the certificates were the crown; friends could come and wear the crown for a while, but the kingdom — that whole ledger of certificates, and the love that bore them — belonged to me.

The first time I went to therapy I was a sophomore in college. I didn't think I needed to be there — I had a mother who looked like Grace Kelly and a father who adored me — but my roommate called the director of the counseling center and insisted he see me. My parents were getting divorced and each of them had begun calling several times a day, crying.

I sat in Fred Davis's office while he asked me how I felt about being placed in the middle of my parents. I said fine. He asked how I felt about my parents getting divorced and I said relieved. He asked if I'd ever been hit and I said yes. He asked if I was in touch with my feelings and I said what feelings. The idea that there was a life beneath the life I was living and that it affected me on a daily basis seemed ridiculous, a bunch of hooey.

The second time I went to therapy I was twenty-two years old. Every cool person I knew was going to see the same confrontational gestalt therapist, so I decided to go, too. During our first session he asked about my childhood and I told him what I remembered — my mother was miserable, my father left when my mother got angry, neither of were ever home. He asked if I was ; I said yes. He asked where my tears

33

were; I didn't know. At the end of the session, he told me he'd never met anyone so out of touch with her feelings. He told me I'd chunked off pieces of myself and frozen them. I never went back. I didn't understand why anyone would rummage around in frozen garbage for the purpose of finding tears. It seemed masochistic, self-indulgent, a waste of money.

The third time I saw a therapist, I was thirty-two and stayed seven years, almost all of which was spent remembering the parts of my relationship with my mother I'd never wanted to revisit. At the beginning of the eighth year, my therapist told me it was time to begin focusing on my feelings about my father, and that until I did, I would never be whole. She said it might take two or three more years of therapy. I quit that day and never saw her again.

I call my father once a week. One ring, two rings, three. He is always happy to hear from me. He is like a welcome committee that never tires of making tuna-and-cream-of-mushroom-soup casseroles even though I've lived in the same neighborhood for forty-eight years.

"Hi, Papa."

"How are you, sweetheart?"

"Fine. But what about you?"

"I'm okay. I keep remembering the first

day you went to nursery school with what's-her-name? That tall red-haired teacher?"

"Mrs. Bolendonk, Dad."

"Yeah, Mrs. Bolendonk. I stood on the sidewalk watching you walk away with your ponytail bouncing. You were so proud of that ponytail."

"I'm glad you noticed that."

"I will always notice those things, sweetheart. You've had my heart forever and forever and that's the way it will always be."

When we go shopping together, which he loves to do, he teases salesgirls, tap-dances in the aisles of stores, sings show tunes walking down Madison Avenue (his favorite is the Richard Burton song at the end of *Camelot*). He always asks the person waiting on us if she thinks we look alike. He loves that our faces are the same shape, our lips the same lips.

During my last visit to New York, he bought me a gray hand-knit sweater and a brown suede vest. While he was standing at the cash register, he asked the pregnant saleswoman when she was due to give birth.

"Next month," she answered.

My father, whose name is Bernard, said, "Do you know if it's a boy or girl?"

"Boy," she replied.

With a deadpan face he said, "I'll give you a hundred bucks if you name him Bernie." After a moment of stunned silence, the saleswoman laughed so hard she had difficulty making change.

We left the shop, held hands as we walked down Madison Avenue. We bought pretzels at a street vendor on 56th Street. I picked off the salt. It was a balmy October day in Manhattan and I felt like I was eight years old again — my dad and I were together and nothing could harm me.

My father was the light to my mother's dark. She was a fat kid who never had a boyfriend until she met my father on a sunny day in March on the quadrangle of Long Island University. Having grown up in the shadow of an older sister who was thin, gorgeous, and wild, she had taken to being the good one, the quiet one, the one her mother didn't notice. At nineteen years old, she had a black miniature dachshund named Widget; blond, straight, chin-length hair; and no self-confidence.

My father's childhood memories are spotty: A pair of twin boys followed two years after his birth, and his mother, never patient to begin with, became distraught. His father, Mark, took a night job in the post office to get away from his wife, and she took to riding the subways all day and night to get away from the kids, whom she left in the care of my father. On that sunny day in March, he was four years back from fighting the Germans in the Second World War, and bristling with bravado. With a toothy smile and a slim, athletic body (he was a regional Ping-Pong champion and could beat the pants off anyone at tennis), he was on the lookout for a voluptuous blond. He didn't have to look farther than a dry patch of grass in front of the student union.

She quit college and married him. They had no money, no help from family. He worked folding pants at Alexander's Department Store during the day, went to law school at night. She had always feared being alone and suddenly every minute was like waking up in the middle of the night — lonely, fearful, shatteringly silent. Within a year, she gave birth to a baby girl — me. By the time my brother was born three years later, my father had graduated law school and was working fourteen hours a day at his new job. She wanted more from her life than a husband who left her alone and two crying babies who needed her constant attention. She lost weight and realized she was beautiful, more beautiful than her sister, who had just dropped her daughter in a boarding house on the way to New Orleans to live a bohemian life. It was my mother's turn now. For the first time in her life, she had money to buy clothes and Cadillacs; she wanted the life she'd never lived — and nothing, not her husband, not her children, was going to stop her.

★ ★ ★

I often get letters from people who tell me they've had a happy childhood and yet are addicted to food, alcohol, drugs, sex. Since they cannot trace their behavior to a particular person or event, they can't figure out why they are so desperately unhappy.

Although not every present-day pattern in our lives can be traced back to our childhoods, the imprint for love — who and how we love, and what we recognize as love — can. To some people love means being left, being anxious, being constantly on edge, and this pattern plays out with frustrating consistency throughout their relationships. To others love means being wanted, being seen, being cherished — and their relationships reflect exactly that. Our earliest experiences of being known or ignored, being held or left alone, being welcomed or criticized, being told we were too much or not enough, create the architecture for love in our nervous systems and limbic brains (the part of our brains that is responsible for attachment) and affect us for the rest of our lives.

Even when a child is wanted and loved, conditions may exist that make it impossible for that child to feel cherished: sickness, poverty, war, family crises. In addition, a mother can want a child without realizing what is actually required of her. The idea of a child and having a child are worlds apart. As Anne

Lamott, author of *Operating Instructions*, says: "Having a baby is like baby-sitting in the twilight zone." You keep waiting for the parents to come home and relieve you of your duties, until you finally realize this is it, you're on your own.

My mother was a goddess to me, a heavenly concoction of fairy dust and golden hair, and I wanted to breathe her into my skin. Everywhere we went people stared at my mother. They thought she was a movie star. They asked for her autograph.

I would have turned myself into a field of daisies, flattened myself into a sidewalk, become a Chinese contortionist with elastic legs on my head to get her love. But just my being alive, just my taking up space seemed to stick out and sting her, make her shrink from my touch. I spoke in words as clumsy as couches, angered her with simple needs. If she caught me looking at her, she demanded I stop staring. When she reached her limit — which changed from day to day — she threw me into my room, and slammed the door. She acted like she was stuck in a house where she didn't want to be, trapped in a family she wanted to leave.

I learned to sense her mood by the way her mouth looked, her eyes narrowed, the timing of her breaths, and I adjusted myself accordingly. If I could figure out what I did wrong before she said it, I could do it right. If I made her unhappiness happen, I could make it unhappen. If I chirped enough and didn't get in her way, maybe she would be glad to see me. I never stopped spilling myself like cake batter into a thousand different molds to make her smile, make her happy, make her love me, and I never stopped believing it was my fault that she didn't.

Underneath the frantic efforts at being bright, being shiny, being good was the belief that my mother was unhappy because she had to look at me, talk to me, be with me. I was the cause, I was the reason. The very fact that I was alive — something in my cells, something in the way I talked, moved, needed, wanted, laughed, reached, slept, ate — was damaged. Was wrong. Was badness itself.

This implicit sense of badness feels more like me than my voice. It is the internal atmosphere in which everything else unfolds, the screen on which the movies of my life are played. No matter what I pile on top — love, friendship, work, velvety clothes, pale pink roses — after a few nights of not sleeping well — illness, overwork, a rejection of some kind, anxiety about the world situation, Matt's absence — it all suddenly dissolves into a chaotic jumble. Matt is a stranger, my house is a foreign country, and I have never done anything good because badness cannot produce goodness. In this place, love is, and has always been, unreal, as if I've hovered outside myself for eons, pretending to have something I know isn't mine, but now I am back where I belong.

The world outside has disappeared into the world inside, but Blanche is a glimmer at the edge of the field. I pick him up, and my heart regulates itself with his thrumming purr. The sheer corporeality of him is testimony to now, here, today. Until I feel his weight on my chest, his whiskers on my face, I don't realize I've left my body. As his edges meet mine, I am drawn back into the present: first my legs, arms, face, then the recognition that I am me and it is today and there is goodness in the world and I am part of it.

The irresistible fact of him — the pampas grass tail, the pastel blue crossed eyes, the way he puddles himself on floors, laps, couches — creates a place for me here on earth, just as my father did before him.

Three

A Zen teacher once said that life is like getting into a boat that is just about to sink. Death: it's the fly in the soup.

My mother says she is afraid of dying, not death. Not me: I am afraid of both. I am convinced that when I die, someone, I don't know who, is going to be there with a huge book, and go through my entire life with me, line by line: the times I told Linda Lang I would poke her in the eye with a straight pin, the times I stole baloney from the Safeway in Buffalo, the homeless people I've walked by without seeing, the wives of the married men I had affairs with thirty years ago, the times I've stood by my answering machine, listening as someone asked me to pick up, all the times I took the biggest piece of cake or twisted the truth. I believe I am getting away with it now but death will bring the final reckoning. Matt will go to heaven. He is courageous and generous, like Meryl Streep in *Defending Your Life*. I will go to hell; I hoard chocolate and believe I am what I wear.

I realize there are only two choices: either I will die before those I love or they will die before me, but I am still terrified. And just as the women in my workshops eat to store for the hunger to come, I store phone calls, trinkets, and paraphernalia-in-bulk to prepare for the moment I am always afraid will arrive tomorrow.

People who try to leave messages on my answering machine now are often cut off because the dozens of messages I've saved from my father leave about three seconds for anyone else to speak. And decades ago, when I showed up in my dorm room the first day of college, I pulled out a two-foot wooden statue of John F. Kennedy my father had given me ten years before. The woodcarver's hand was not as steady as it should have been, which left Mr. Kennedy with disproportionately large front teeth and feet the size of small canoes. My roommate, Jace, insisted it was the ugliest thing she'd ever seen. But that was before I pulled out the two white ceramic rabbits, also gifts from my father, which she said looked capable of devouring small children. Until she underscored their aesthetic liabilities, I'd never really assessed these gifts; I only knew that since my father could die at any time, I had to save every bit of concrete proof that I was loved before he vanished from the earth. The things he gave me were my certificates of love, my diplomas, my credibility.

As I've gotten older, my death fears manifest in ever more dazzling and creative scenarios. Each time Matt is about to leave on a road trip, before he gets in the car, I feel the need to apologize for everything I've ever done, every name I've ever called him, in case I never see him again. Since my behavior is frequently unruly, this ritual makes parting a cumbersome event. If he is in a rush, I often end up outside his car, in my nightgown, screaming out things like ". . . and I'm sorry I ate the cashew chicken salad yesterday, which I pretended you bought for me but I really knew you wanted for your own lunch . . ." Any lie, anything I've held back, any way I've loved him but haven't mentioned, I believe I have to say *now*.

But it is my fears of Blanche's death that qualify me for a new definition of insanity. When he sheds whiskers, I save them in a black-and-white-checked mother-of-pearl box, along with tufts of his curly stomach fur and butterscotch wisps of his tail. And ever since a friend mentioned that she might be able to make a sweater from cat hair, I've been saving bags and bags of the fur he sheds when I brush him every night, complete with dead fleas. I take thousands of photographs of him, tape his morning meows on my computer, and carry an alarm clock with his amplified purr as the wake-up tone.

In an attempt to reassure myself that I am

unnecessarily concerned about Blanche's death, I consult a team of animal psychics. Each of the pet communicators uses completely different methods of contacting Blanche.

Todd, an erstwhile lawyer, needs to have him in the same room, talks to him out loud, and then tells me, sentence by sentence, what Blanche says. Alana works over the phone. She sets the receiver down, goes off for about fifteen minutes to speak with Blanche on the astral plane, then comes back and gives me a full account of his thoughts and responses. Ingrid seems to be especially attuned to the sphere of animal supplements, since after telling me about Blanche's state of mind, she recommends many expensive vitamins, all of which she happens to sell.

Through Alana and Todd, I learn that Blanche is content with his life, his body, his environment, and that he has acid indigestion. Neither of them helps allay my fear of his death, though once, when Blanche was waking us up every morning at five a.m., Todd asked him what was going on. Blanche said he was bored because I was writing all the time and wasn't playing enough with him, and that waking us up early was an attempt to get our attention.

Todd suggested I buy some toys for Blanche with tracking balls and motorized mice. After I came home with three bags of gadgets, scratching posts, and the feline equivalent of a motorized toy train, and set aside half an hour a day for quality time with my cat, Blanche stopped waking us up in the mornings. Even Matt, who doesn't believe in psychics, astrologers, synchronicity, past lives, or anything else that smacks of New Age California, was impressed, though he was quick to point out that Blanche's morning silence coincided with his getting stoned every night from the catnip in the whizzing mice.

It's not that I am untouched by death — my first love, a high school boyfriend, died suddenly of a rare kind of cancer when I was sixteen. A few years later, a close friend died in a car accident. All four grandparents, a beloved aunt from childhood, and three friends with AIDS have died in the last decade. But Blanche is different; he anchors me to life.

Three times I think Blanche is lost and call Matt at work. After he drives the two miles home from his office, after he shakes Blanche's food bag outside, and walks up and down the block, for fifteen minutes, thirty minutes, an hour, Blanche finally appears, looking sleepy and confused about why such a fuss is being made when all he was doing was having an afternoon siesta under a neighbor's porch. He looks at me sideways with a bit of disdain, as if to say, "I've been

46

tolerating these hysterics for ten years now. I understand you had a troubled childhood, but really, Geneen, when are you going to realize that being alive is good?"

But I can't move past the fear. Blanche dying is unacceptable. I am caught in my familiar pattern: believing I won't live if he dies. Intellectually, I know this is nonsense — people live through the death of their children for God sakes. Certainly I can survive the death of my cat.

The fourth time I think Blanche is lost, I spend two frantic hours looking for him in the neighborhood, an hour crying on my bed, and when he saunters out of the laundry basket where he has been conked out beneath some shirts, I feel like Dorothy in *The Wizard of Oz* when she realizes that if she ever goes looking for her heart's desire again, she should look in her own backyard. I realize I need to start looking in my backyard heart, where I must have buried whatever I depend on Blanche to fill.

It occurs to me that I can spend the rest of my life (and his) in low-level panic, or I can take a leap into the suffering, and make friends with fear, pain, and sorrow. It is the same juncture I reached with food, when I realized I could keep being frightened of going off the diet and eating so much I'd end up weighing a thousand pounds, or I could stop dieting and discover if there was a bottom to my hunger.

The next week, I take two drastic measures: I commission an artist to immortalize Blanche by painting three portraits of him, and I make a commitment to find a spiritual path whose focus is to help people remember their true nature. I figure it is good to cover all the bases: if I discover that my true nature is nothing to write home about, at least I will have a lot of nice paintings.

Four

I gave up on God about the same time I started dieting. It's not that He (there was no question then about God being a She) was ever a big part of my life. We were bagels-and-lox secular Jews, who never talked religion or faith, and rarely went to temple, even on the High Holy Days. Once, on Yom Kippur, I took the bus with my friend Sharon to Temple Beth-El in Flushing. We were supposed to be fasting, but when we passed Antonio's Pizza Stand on Ditmars Boulevard, we faltered in our resolve and decided to verify what the Scriptures said. We called our local expert, Sharon's mother.

"We're hungry," we chanted. "Does it count if we eat pizza?"

"No," she said. "Go right ahead. Pizza is Italian, and this is a Jewish holiday."

In my mind, God was the person who helped Charlton Heston part the Red Sea in *The Ten Commandments*. God was the old man in the song "It's raining, it's pouring, the old man is snoring." God was also the one who allowed forty members of our extended family and six million other Jews to die in concentration camps. He was not to be counted on for help.

Ann, our housekeeper, talked out loud to God all day long: "God," I'd hear her say, "I don't see why you had to go and do that. Be better if you just tol' people your reasons for things."

One day I walked into the kitchen and heard her telling God about her husband, Wilbur, who was having a difficult time at work with his boss, a mean old man named Roscoe, who yelled all the time.

"Ann," I said. "Why do you do that? He's never going to answer you."

"God has ears on toasters and doors and bedposts and clothes. He hears everyone, all the time. It's talking to Him that matters, Geena girl," she said as she thumped her heart. "Makes me feel good. Don't matter if He never answers."

I began praying in earnest when I was eleven, the year my mother told me she and my father were getting divorced. *Please God, let Mommy and Daddy be happy together.* Every night for a year, I got on my knees just like the pictures of the girls in the books I'd read years before, and asked God to help my parents. I didn't like to beg, but I was willing. This was important; my life depended on it.

My parents didn't get divorced for eight more years, but their misery clung to the air like Agent Orange, invisible, deadly. I couldn't understand who God was if He didn't answer prayers, if He couldn't stop mothers from hitting, fathers from leaving, dishes from being broken. What was the point of gushing, of being painfully vulnerable, when He didn't answer? I didn't want to be a sucker, didn't want to keep getting disappointed, so I washed my hands of the big guy and turned to food instead.

The first time I overate, my suffering about my parents disappeared in the bliss of coffee ice cream. I felt what people who have religious experiences must feel: release, power, grace, surrender. At twelve years old, I knew the giddy freedom of being able to alter my state. Why pray when I could do for myself what God refused to do?

Until my twenties, food was my stand-in for God. I turned to ice cream for the same reasons people turned to Him: to comfort and sustain me, to remind me that sweetness and heaven were possible.

In September of 1975, I was twenty-four years old and working as a counselor in a suicide prevention and crisis agency. My boyfriend, Jack, and I were spending the day on a beach in Carmel. He was reading Peter Beagle's *The Last Unicorn* and was so enthralled with it, he hadn't spoken for two hours, so I decided to take a walk. Halfway down the beach, I met my college friend

Ruby, whom I hadn't seen for more than a year.

Ruby and I were bingeing pals: we ate gallons of Breyer's fudge swirl ice cream with our fingers (mining the veins of chocolate first), we regularly made, then ate, one huge chocolate chip cookie from a recipe intended for thirty-six, and we could slurp three-dozen oysters each in a couple of minutes. We were unbridled in our attempts to break free from restrictions: we had tried all sorts of sexual combinations, experimented with drugs, and painted our cars, walls, and clothes in splashy jungle motifs. The Ruby I'd known was wild, throbbing, ablaze, but the Ruby I met on the beach was smooth and settled, as if instead of ice cream and chocolate, she'd eaten turquoise sky.

She said, "I just came back from India."

I'd never known anyone who'd even wanted to go to India; it seemed so exotic, so far away, so full of malaria.

"Wowsers!" I blurted, after which I felt profoundly unholy. And then, "What did you do there?"

"I found the Godhead," Ruby said.

The conversation had now wandered so far from chocolate chip cookies, I felt as if I were hallucinating. I had no idea what to say next.

"Just the head?" I asked.

"No, silly. 'Godhead' is just an expression.

It means God. I found a living embodiment of God. He's called an avatar."

"Time to breathe," I said to myself. "She must have been abducted by a cult. Maybe they are lurking behind the rocks. She might need my help."

I asked her to sit down and tell me everything there was to know about godheads.

After an hour of conversation, the person who made her way back to Jack was totally different from the one who'd left him. He was in exactly the same position, spellbound with unicorns and magicians, unaware that my world had turned upside down.

"I'm quitting my job and going to India," I said.

"And I have blue hair," he replied, without looking up.

But the time I'd spent with Ruby convinced me that she had learned something I needed to know. I was unsure about the God part, but I wanted the peace I saw floating behind her eyes, rippling across her face. I wanted blue sky in my chest, belly, heart.

Robert Frost wrote that a poem "begins as a lump in the throat, a sense of wrong, a homesickness, a love sickness. It is never a thought to begin with." After speaking with Ruby, I was homesick for a place I had no recollection of ever visiting. I knew it wasn't a physical location, but since I had no idea where to look, and she believed it was in

India, I figured I might as well start there.

Three weeks after my chance meeting with Ruby, I'd quit my job, sold my car, and was standing in my mother's foyer, packing to leave for Bombay. My father, who lived in an apartment nearby, stomped through the front door after work. His navy-blue-and-silver striped tie was open at the neck, his face was gray and mottled. I smelled liquor on his breath and realized I'd never seen him drunk before.

"You're not going to India," he said, slurring his words.

"Oh, Dad, I'm sorry you're upset, but it's a done deal. I'm going." I couldn't believe these words were coming out of my mouth. When he had wanted me to get straight A's, I did. When he had wanted me to graduate Phi Beta Kappa, I got my gold key, walked straight to where he sat, and handed it to him. When he had forbidden me to mention my stepfather's name in his presence, I didn't argue with him. But this time I was unfazed by his protests. The inarticulate longing drawing me to India was even stronger than my desire to please him, and although I was terrified about wandering alone in the rat-infested streets of Calcutta and watching bodies burn on the Ganges, I knew I had to go.

I'd already been through a milder version of this conversation with my mother at least fifty times. She said: "Most people's children go to France, Italy, England. Why do you have to go halfway around the world? What are you looking for that you can't find in Paris or New York City?"

I couldn't answer her. I didn't dare tell her about Ruby. As a child, when she said no to something I wanted to do, and I whined: "Mary's parents said she could go, why can't I?" she answered, "I don't care what the other kids do, it's you we are talking about." Going to India was the ultimate "If Ruby can do it, I can, too." But it was true — I believed if Ruby could have blue sky in her chest, I could, too.

My father jerked his tie off his neck. He was frustrated, furious, shocked at my defection. He shouted, "I am going to call the INS and have you declared an illegal alien. I will do anything to stop you from getting on that plane." He stood up, glared at me, and stormed out of the house.

Five hours later, I was sitting on an Air India flight to Bombay.

For three weeks I traveled by train and bus to Calcutta, Agra, Jaipur. I got sick with dysentery and spent three days throwing up. I watched bodies burn on the Ganges and touched the hem of an Indian saint called Anandamayama. Then I went to southern India and met the man Ruby called God. He had frizzy hair and stained red teeth from chewing a plant called betel. He didn't look or walk or talk like God, but I had flown seven thousand miles to see him, and I was willing to give him the benefit of the doubt. Every day for three months, I spent five hours at his ashram, meditating, chanting, watching him walk, listening to him talk.

I didn't have a born-again conversion — in fact, the avatar never addressed me directly except to tell me to be quiet — but I did have blazes of moments during which I felt utterly different from my old familiar self. Though they seemed to happen randomly — when the scrabbling of the cockroaches woke me up at two a.m., when my sari was flapping in the wind as I rode my bicycle, when I was learning to meditate and couldn't concentrate for longer than three seconds — I understood that even if the avatar wasn't God, even if God didn't exist, there was a vast mysterious fabric woven through everything, including me. When I felt connected to it, I felt more like me than I'd ever felt and more at home than I'd ever been in my life. My fears about death seemed to disappear. The contrast to my familiar self was stunning. It was as if I'd spent my life gravity-bound and landlocked, wheezing along as a banana slug, only to discover that I was a ruby-throated hummingbird.

By the time I stepped off the plane in New York, I couldn't go back to believing that being the richest, thinnest, best-dressed banana slug around was the best I could do, but I still had no idea how to live in the sky.

Over the next eighteen years, I attended dozens of Buddhist meditation retreats and learned to follow my breath, spend days in silence, sit quietly for hours at a time. Soon after we met, Matt and I took a monthlong retreat with Vietnamese Buddhist teacher Thich Nhat Hanh at his center in the south of France.

During the retreats — and for a short time afterward — I felt as if I'd entered a mysterious, shimmering web of aliveness; I felt spacious and luminous, clear and confident, but it never lasted. It seemed as if I was magnetized to a particular way of seeing myself — as doomed, as lacking, as always missing the mark — and no matter how much therapy, how many expanded experiences I had of being vast and loving and open, I slipped back into my Geneen costume again. It was like wearing the same clothes for thirty-five years; the buttons were popping, my arms were twice as long as the sleeves, but it felt like a second skin and I couldn't throw it away. The familiar fears of death, of loss, of believing that any other life than this was not possible, or if it was, it was for spiritually advanced people and not for me, came clunking back into my bones.

The Sufi poet Rumi said of God: "Ever since I heard my first love song, I began looking for you." Despite many years of retreats and meditation practice, I'd never stopped looking or wanting to find this you, but since I still didn't believe in the God with white hair and suspenders, and since I still found this persistent longing confusing and embarrassing, I felt hopeless about ever discovering it/Him/Her. Also, in my mind, true seekers lived on spare diets and spare budgets, in brown polyester clothes, sturdy shoes, and no lipstick or blush, and since brown made my face look muddy, and living without my black zip-up ankle boots was nonnegotiable, I concluded I was a fake.

Nineteen years after my trip to India, my new friend Taj tells me about the inner-work school in which she is a student. She says they focus on being in your body and using your present experience as a starting point for inquiring into your true nature. I've tried and been disappointed in so many of these spiritual practices that I am cynical about launching into another one. But Taj is the first friend with whom I can roam in the same conversation from the snappy black pants with the frilly bottoms at Molly B's to the longing and homesickness I've felt for years. I sense she, like Ruby, knows something I don't about working with fear and being at peace. I sign up for my first eight-day retreat at the school.

Stephen Levine says that hell is wanting to be somewhere other than where we are; I am in hell for the first three days of the retreat. On the fourth day, it occurs to me that the worst that can happen is that I will stay for the full eight days, not learn anything, and my life will be as it was before: having enough and acting as if I have nothing; living in constant terror of death; believing that there is something damaged and irreparable at my core. I decide to stay.

On the fifth day of the retreat, I stop feeling like they are going to shave my head, take away my ankle boots, and make me wear brown muumuus. I begin to understand that the main practice of the school — inquiry into the present experience — is an expanded version of the one I use for addictions, only theirs is an ongoing, daily practice, and I stopped doing mine years ago when eating lost its glamour. By integrating the sophistication of Western psychology — including the work of Freud, Jung, and recent child-development theories — with religions and Eastern wisdom traditions — Christianity, Judaism, Islam, Buddhism, Hinduism, Taoism — the school teaches that there is no need to reject any parts of yourself, even the ones you think are selfish or mean. Just as I believe emotional eating is the external expression of deeply held beliefs about self-worth and body size, they believe that the ways we feel stuck and unhappy are doorways to decisions, judgments, and beliefs we made long before we knew what we were doing. And if we don't reject them, if we are curious about them, if we learn to soften to the pain of a lifetime, we will open to what was there before all those patterns developed: our true nature, which goes by hundreds of different names — Christ consciousness, Buddha nature, Spirit, essence, Atman — depending on the tradition.

By the eighth day, I experience similar bursts of the shimmering I'd felt in India: moments when time opens up and everywhere I look — the grass, the sky, my hands, a desk, a face — is a feast. During the inquiry sessions, I notice that beliefs I've been reacting to for years — that I am unlovable, too intense, damaged — become translucent, like an opaque veil that suddenly turns diaphanous.

This isn't the me I call myself, and I want to bottle it immediately so that I can take it home and pull it out when I am miserable. I do the next best thing: I sign up as a student in the school.

Five

We celebrate Blanche's twelfth birthday by giving him a bowl of sardines and corn bits that Matt nibbles off the cob. We try to teach Blanche to hold the cob between his paws, but he can't maintain the effort necessary to keep the cob aloft.

Blanche begins seeing a cat chiropractor every two months so that when he jumps down from the kitchen sink after taking his thrice daily drinks, the pounding of his girth upon the wood floor will not jolt and misalign his vertebrae. He also continues receiving acupuncture, Chinese herbs, and homeopathic remedies, while I continue my own version of being in alignment: meditating, attending retreats, and having weekly sessions with my new teacher, Jeanne, from the inner-work school.

Slowly, very slowly, I begin to experience days and weeks of feeling the kind of happiness I'd previously associated with being on drugs, being with Blanche, or being in love. It dawns on me that the answer isn't in Blanche, it's in the way he makes me *feel* — light, content, joyful — and that these are birthrights, not just unrepeatable accidents.

Up to this moment, my life has been a series of attachments to the mirrors (i.e., animals, people, situations) that reflect qualities of sweetness, love, and unapologetic joy, instead of recognizing myself *as* them.

In people years, Blanche is now eighty-four years old; my father is seventy-six. Since no one in my family has died before eighty-seven, I am sure that of the two of them, Blanche will go first.

The day I hear my father is ill starts as an ordinary day. I write, spend time with Matt, take a walk, pay bills. At five p.m., with coat on, keys in hand, I call my father to say I am leaving for a weeklong writing retreat and will speak to him when I return. His wife, Pepper, answers, tells me he hasn't eaten for three days, and just like that, everything changes.

Two days later, after we cancel the retreat and arrange a series of diagnostic tests, my brother Howard and I speak to my father's doctor. "The radiologist found the most interesting thing," the doctor says. "It's fascinating! He's never seen anything like it in his life! Two HUGE masses on each of the adrenals. So huge, they must be six inches each. We are amazed."

It doesn't make sense, hearing the words "fascinating" and "interesting" with "masses" and "tumors." I am so confused I can't think.

My brother says, "We are laypeople. Can you explain?"

The doctor replies that the masses are either metastatic illness or lymphoma, but almost definitely not benign. He says my father needs to have a biopsy done, some further tests. He never mentions the word "cancer," never stops talking, keeps moving like an icy river ripping up trees and boulders and any other live thing in its path. I can't feel anything except my hammering heart and the pounding in my ears and the vague sense that this man has just said without saying that my father is dying.

A few days later, we speak with a doctor at Sloan-Kettering, who tells us that in addition to the tumors on his adrenal glands, my father has stage four lymphoma in his stomach and that his chance of living more than a year is less than thirty percent. She says he needs to go into the hospital *today;* otherwise his calcium levels will rise so high that he will lapse into a coma and die within a few days. She also says that he should begin chemotherapy immediately.

Last Tuesday, I didn't know he had cancer.

Next Tuesday, my father could be dead.

I scour libraries and bookstores for guidance; I find many different kinds of books about death and loss, but none of them speaks to the raw primal terror I am experiencing. No one seems as desperate or wild as me.

I want to scream at time to stop. I want to scream at everyone I see, beg them to STOP EVERYTHING and help my father.

Nothing else matters. (Was it really yesterday that I cared about which shade of cream to paint the bedroom?)

Pepper is my father's fourth wife. They met when she worked as an underwriter in his mortgage company while he was still married to my mother. She was twenty-seven; he was forty-two. After his divorce from Betsy, my father's third wife, he lived with Pepper for ten years before they married. My brother was supposed to be the best man in their wedding, but he got stuck in traffic and missed the ceremony.

My father treated Pepper as an underling who adored him, and whom he finally decided to marry because she'd been hanging around for twenty years. He treated her as his sure thing, his safety net. Someone who would care for him when he got old.

He was wrong.

Pepper works every day from eight-thirty to five-thirty. She never learned to drive, and my father says she is frightened of doctors. In the first week after his diagnosis, my dad has to go to the doctor twice for tests, and Pepper says she will get fired if she takes off work. She says he can take a cab.

My father has Parkinson's disease, and his steps are uncertain and wobbly. I feel panicked and unbearably sad at the idea of his going to the doctor alone. I want to beam myself to his side, tell him I love him, and while I am there, I want to strangle Pepper.

I tell myself that rage is a reaction to pain, and it is easier to blame Pepper for his illness than to feel the searing grief. But I still want to strangle her. Why didn't she get alarmed when my dad lost thirty pounds in three months? Why, when he hadn't eaten or slept for three days, did it take phone calls between my brother and me to get him to a doctor immediately? Why does she constantly yell at him to pick up his feet?

My dad says, "She can't take off work. She will be fired."

"What boss is going to fire an employee for taking her husband to get a biopsy?" I ask. I can't tell whether Pepper is lying or my father is covering for her, but I know I am hearing partial truths.

Then he mentions that she doesn't spend her work money on anything but clothes, and that since she has gained so much weight in the past few years, she needs to keep buying newer and bigger sizes. He says there is no more room in the closet for her stuff, and so her blouses and pants hang over every piece of furniture. "That is why I tripped last week and fell on my head," he explains. "I got my sweater caught on a hanger sticking out from a chair."

"Is this supposed to make me feel better?" I ask him, outraged.

"Don't be so hard on her," he says. "She had a tough life."

My mother divorced my father twenty-nine years ago and married a man named Richard Wiggs. My dad, by his own admission, never stopped loving my mother. He'd made dozens of cassette tapes for her of love-lost songs, by Frank Sinatra, Eydie Gorme, Lanie Kazan. He'd sent her flowers every year on October first, their wedding anniversary. Every time she walked into a room, the sun came out on his face. When he married Pepper, he said to my mother, "I'm tired of waiting for Wiggs to die."

She and Dick moved to Florida five years ago and coincidentally moved back to New York the day of my father's diagnosis.

When I call her panicked about my dad going to the doctor alone, she says she will take him. She promises she will not let him take cabs to appointments and that she will spend time with him, be gentle with him, make sure he eats.

I was furious with her for thirty years. Now

66

she is putting her life aside to help my father, brother, and me. She is being the mother I always dreamed of having. It's strange how things come around.

She says, "I am doing this for you, not for him. I would stand in traffic for you and your brother."

I giggle when she says this; understatement was never one of her most striking features. We have that in common.

Soon after my father was diagnosed with lymphoma, I was invited to hear a spiritual teacher named Byron Katie speak. She asked people to call out their judgments and opinions. I said, "My stepmother should take better care of my dying father." That seemed obvious: when someone is ill, especially your husband, you put everything aside. Katie turned to me and said, "Let's work with you, sweetheart."

Okay.

She asked me if the statement I made was true.

Yes, of course, I said, it is true.

But is it really true? she asked. And how can I be so sure? Do I really know what Pepper's destiny is, what the spoken or unspoken contract between her and my father is? "Well, no," I admitted, "I don't." I can't possibly know what she is supposed to be doing, I can only know what *I think* she is supposed to be doing. Maybe, Katie pointed out, she is a smart woman. She is letting her husband live his own life.

Then she asked if there is any way that holding on to this belief is not causing me pain?

No. Every time I think of Pepper not taking my dad to the doctor, I want to kill her. That definitely classifies as pain.

Who would I be in her presence without this belief?

I'd be me in pain at my father's illness being with her in pain about her husband. Two beings who are facing the possibility of losing someone they love.

Katie pointed out that letting go of my judgments about Pepper frees me.

But did I want to be free? Or would I rather stick to my position about Pepper and get everyone to agree that I am a long-suffering daughter with a wicked stepmother.

Uh-oh. I wish I could say I'd rather be free, but I am very fond of my positions. And as much as it hurts, I am also very fond of suffering. It gives me something to talk about and a way to connect with other people. I believe that if I am suffering, they cannot possibly dislike me or be threatened by me. If I am suffering, they will love me. No wonder I am so attached to pain.

A student once asked Zen master Shunryu Suzuki Roshi what nirvana was and he said,

"Seeing one thing through all the way to the end." That night I make a decision: I will see my father and myself through this to the end without my usual dodging maneuvers: blaming someone else, blaming myself, or resorting to my Ph.D. in emotional eating.

I soon realize there is something else here besides grief, and it is so surprising that I keep turning away from it because the impending death of someone you love is not supposed to feel like this. In addition to my usual thrashing, I am aware of an unchanged and unchangeable stillness, a peacefulness so languid and so relaxing that it feels like being dissolved, cell by cell. The difference between how I think death should feel and how it does feel is like the difference between being knocked around on the surface of the ocean and gliding along on the dark, hushed bottom, where the surface is only a dream.

When I pay attention to what I am losing, I feel insane with grief. When I pay attention to the enveloping stillness, I know that nothing bad is happening.

It makes me wonder if what the Sufis have said for a thousand years is actually true: anything you can lose in a shipwreck is not really yours.

I once heard a talk given by a Tibetan Buddhist teacher whose son had died in a car accident. For one week after his death, she did all the things a good Tibetan Buddhist would — prayers, mantras, ceremonies — but they didn't help the grief. Then she realized all those things were keeping her from facing the fact that she was never going to see her son's face again. She keened and wept for five days.

She said that when people die, it seems as if the pain is almost unbearable, but only almost. She said if you really feel pain, it washes you clean and you can actually feel joy again. She said most people never go to the bottom of their grief because they are too frightened of drowning in it. She also said that unless you feel it all the way through, it haunts you in bits and pieces forever.

I liked her because she was wise. I also liked her because she had been watching *All My Children* for twenty-five years, like me. It takes great courage to be a spiritual teacher and admit you watch soap operas.

On the day before I fly to New York to see my father, Blanche breaks a bone in his right foot jumping down from the sink. When I tell this to my friend Judah, she says that when people were grieving hundreds of years ago, they would break parts of their bodies so their despair inside would show on the outside. Judah says that animals do the same thing for people. She says that maybe

Blanche broke his foot to take on some of my pain about my father. Even though Judah lives in Massachusetts, I know that if I told my friends in New York what she said, they would roll their eyes and say, "Only in California."

I see my dad the first day he goes into the hospital. He is waving his arms, muttering. He doesn't recognize me. I begin to cry the second I see him. Howard says that as soon as his calcium levels go down, he will be coherent again.

I am stunned by the acrid smell, the three other men in his room (one of them cries "Help me! Help me!" every two minutes), the shrill, continuous beeping of IVs, the overworked, bored looks of the nurses. Breathing feels like slogging through quicksand; the oxygen in the hospital is stale, as if it's already been used, and after an hour, I am more tired than I have ever been in my life.

Nothing here is familiar, least of all this person dressed in my clothes, talking with my voice, standing at the side of an old man's bed. As I stand there, I swing between being an adult I don't recognize — who knows what to do, is not afraid of death — to feeling like the frightened child who used to pad to his side of the bed and crawl in next to him until I fell asleep.

71

Six weeks ago, he looked like my father. Now he looks like a stranger. I recognize the shape of his nose, his mouth, his hands, but the man I have called Papa is not here.

I talk to him.

I sing to him.

I say, "You are surrounded by light and love, Papa, can you feel it, can you see it?"

And he whispers, "I'm trying. I'm trying."

I sit with him for ten hours.

I forget to eat, drink, take walks outside.

I've never been with anyone who is dying, so I don't know if this is what death looks like.

He didn't prepare. He never believed death would come; he put his faith in his collections of Mont Blanc pens, antique watches, Paul Stuart suits. He built his life on things that didn't last, couldn't keep him company now. Although he didn't talk about religion, clearly he thought it was for sissies; belief in God didn't keep forty members of his family from getting gassed at Auschwitz.

Still, with no faith in the invisible world, all that is left for him to rely on is what he can see, touch, or buy, which today, in room 409 of Long Island Jewish Hospital, is nothing at all.

My chest feels like shattered glass. Even the air is wounded.

After a few days in the hospital, my father recognizes my brother and me.

He says, "I almost died."

I say, "I am so glad you didn't. Were you scared?"

He pretends he doesn't hear me and looks away.

Still, I am overjoyed to have him again. When I touch him, listen to him talk, look into his eyes, I try to memorize the map of his arms, his smell, the texture of his skin. Right now, it's enough that he's above ground. It's more than enough. It's everything.

The nurse tells my father he has to eat, otherwise he won't have the strength to get better. He takes the tiniest speck of baked salmon on his fork and then puts it down. "I can't," he says, "I just can't."

Rice pudding was my father's favorite dessert. He also loved bread and butter and mashed potatoes and creamed spinach. He could eat a whole row of Fig Newtons in ten minutes. My father was once overweight and I worried that he would have a heart attack; when he ordered rice pudding, I would tell him he shouldn't. Now his fingers are so bony that his rings fall off, and his back is so frail that it feels like a sparrow's leg. If he wanted rice pudding, I would bake it myself.

Every day, when I leave the hospital, I go to the Silver Moon Diner on Union Turnpike, where my dad and I used to spend hours talking in the red leather booths when I was twenty-four years old and anorexic. Since I never ordered anything, he'd say to the waitress, "She doesn't eat much, but don't worry, I'm a big tipper."

In those days I weighed eighty pounds. I didn't know what to do with my life — I was working as a maid and dishwasher in Big Sur, and thought that being thin was a good start. An hour before my father was scheduled to get on the plane after visiting me for the weekend, I told him I wanted to kill myself. He said, "Get out of Big Sur and find something you love to do, I don't care how much money it costs." He said he would support me for as long as I needed. Every day after that, he called and urged me to get going. Within a month, I moved to Santa Cruz. Within three months, I began to write poetry and make avocado and cheese sandwiches in a health food store to support myself, and he didn't need to pay my rent anymore.

Sitting in the diner booth now, with the towers of the hospital in view, I feel wildly grateful to have known him, to be alive.

Thousands of people in my workshops tell me they hate getting hungry because they are afraid to overeat. They hate their appetites.

They hate their bodies. If they saw my dad now, they would change their minds. I feel like papering the sky, writing a letter to the fifty thousand people on my mailing list, saying: FORGET EVERYTHING ELSE BUT REMEMBER THIS:

If it doesn't taste good, it's not worth eating.

If it doesn't bring you pleasure, it's not worth doing.

Celebrate every one of your hungers.

You will never be sorry.

A student of mine told me that when sacred vessels used to break, they'd mend them with liquid gold. The more they broke, the more valuable they became. The same is true with hearts, she said. Still, it seems important to know when to take a break from broken hearts and go shopping.

I buy a pair of sparkly red mules at the end of a day I spend with my father. They remind me of Dorothy's ruby slippers. Even though she thought she had lost everything, Dorothy never truly lost the power to return to what she loved. I am starved for beauty, and I might be going a little overboard because I buy a pair of green mules, too. It only takes twenty minutes.

Walking into a store's extravagance of color and texture and buzz is like food to me now.

I think I've discovered a stage of grief unknown to Elisabeth Kübler-Ross: death-and-dying retail therapy.

I keep telling myself that I am out of control and have to stop shopping immediately, but I don't. Then I begin to worry about going bankrupt the same way I used to worry about getting fat, and my pain at my father's illness is blotted out by my worries about having to sell everything to pay for my next three thousand pairs of earrings. But *then* I decide I will write a book called *What to Eat and Where to Shop When Someone You Love Is Dying*, which makes my forays to the Miracle Mile scholarly research.

I buy a heart-shaped belt buckle on my latest shopping trip. When I was twelve, my father bought my brother a movie camera and I cried because he didn't buy me one. Then he kneeled down next to me and pulled out a little gold heart with rubies in the middle. He said, "This is my heart. You have my heart. That is as important as any movie camera." Since then, I have been a sucker for hearts. Soap dishes, dinner plates, pillows, earrings, tables, cutting boards, rocks, wreaths. In my first apartment, I decorated my bathroom with frilly cardboard Valentine heart-boxes from Walgreens and was deeply offended when a friend called it "beyond tacky." The truth is, I have always been blind where my father is concerned.

After my father has been in the hospital for three weeks, the doctors say he can go home the day after Christmas. In the meantime, they say, enjoy the holiday.

Five days later, I kiss my father good-bye before flying home to California. Each moment feels suspended, as if we are moving through water: his forehead on my lips, his bony hand in mine, the smell of Old Spice aftershave. I put my head on his chest, tell him I love him, and that I will return in ten days. "Good-bye, Ollie," he says as I am walking out the door. I turn to look at him and he is waving his right hand in the shape of a rainbow, the way he did when I was five.

At home, I carry Blanche everywhere. When he isn't in my arms, and I am not crying in Matt's arms, I stand at the kitchen sink weeping. I come out to my studio, meditate, write, teach, talk to friends, call the painter, answer e-mails, and all the while my father is dying. On the surface, nothing has changed. He was never a part of my life in California. But he does not live or die on the surface of my life; he is the bottom of everything. He is the invisible container in which my life unfolds.

I have never lived a day of my life without knowing he was alive, and I am terrified of

77

losing him. Night after night I wake up sobbing, thinking, "Don't leave, Daddy, don't leave. You promised you wouldn't leave." Sometimes I wake up Matt. Sometimes I pad around the house, put on lights. Sometimes I think middle-of-the-night thoughts: "When he dies, he won't answer the phone anymore. He won't call me 'my darling.' When he dies, I will never hear his voice again. When he dies, what happens to his name in my address book?" During the day, I recognize that voice as the three-year-old clutching her daddy's hand, but in the middle of the night, his death seems unbearable.

Now I know why people go crazy when someone they love is dying: one major loss and a whole lifetime of unspoken sorrow and fear come crashing in.

The first time I see my teacher Jeanne after I return from New York, I feel as if Love has a face and she is shining on me. She reminds me that I can allow myself to feel the pain of my father's dying and not get sidetracked into stories of "what if" and the fear of being destroyed by so much intensity. "But there is shattered glass in my chest," I tell her. She says, "It is only a *feeling* of glass," and asks if I can breathe gently into the sensation, and describe it to her. She says things like, "Is it pushing in or pushing out? Does it have a density? What happens when you focus on it?" Then I realize that it was shattered glass yesterday, but today it's the feeling of being without skin. "Oh," she says, "no skin. What does having no skin feel like?"

Leaning into that rawness is not my first inclination, but when I stay with my direct experience, I begin to feel fluid and non-bounded, as if my molecules are mixing with clear space, as if there is no distinction between me and not me. An openness arises, a calmness.

Now that I have to face the unfaceable, I feel washed clean, unexpectedly alive. Everything I push away comes tearing through the cracks. There is nothing separating me from wild terror or wild freshness. They seem to go together, and at quite a cost: if I give up holding on to what I love most, I get the world back. If I give up trying to orchestrate the safety and the future of those I care about, I get to have everything — as if for the first time: the gratitude of having a body, the round shimmering coins of Matt's eyes, the silkiness of Blanche's fur beneath my fingers, the delicacy of air, just plain air, on my skin. Geez. Who made this bargain?

At a gathering with friends, I weep about my father's death until I am utterly still inside. When I look around the room, I see the tenderness in their eyes, and a knowing arises that this is what is going to get me through.

Not their personal love — my dad loved me and now he's dying — but the fact that Love exists, is real, is true. It existed before and through him and will continue after him. I am beginning to believe that everything will not end when he does.

Matt and I watch *Ally MacBeal*. It is her thirtieth birthday and her dad calls to wish her happiness. As soon as she says, "Hi, Dad," I begin to cry. "I won't have a dad anymore," I say to Matt. "No dad to call me on my birthday, to perfectly remember the day I was born. He's one of two people in the whole universe who knows what I looked like when I was two hours old." Then it hits me that my father wasn't there the day I was born; he was at work. I also remember that he forgot my last three birthdays. I had to call him days later to remind him that he forgot.

The truth is, I won't have a dad to call to remind that it's my birthday. When I say it that way, I feel less sorry for myself.

A week later, I walk into my father's apartment. He is sitting in a chair covered with sheets so that if he pees through his diapers, the chair won't get ruined. He keeps asking what the L word is that he has. He won't even mention the word "cancer." "L-y-m-p-h-o-m-a," I say to him. "It's a form of cancer."

"Oh," he says, as if I just told him he needs a tetanus shot.

I want to talk about things like love and whether he is scared of dying. But every time I begin a conversation, he grabs the remote and switches on the television set. The only thing he wants to do is check the stock market.

I tell Matt about my dad's silence, and he says, "I guess this isn't *Tuesdays with Morrie*."

When my father first got sick, I arrived in his hospital room with yellow Ecuadorian roses, Bach violin sonatas, Rilke and Mary Oliver poetry, audiotapes about learning to be with illness, and family photographs. I was going to create a loving environment, uplift his spirit. I knew he'd spent the last dozen years somewhat depressed about being retired, and I figured that being sick and vulnerable could break through isolation. I was also hoping to share the parts of my life he was never interested in — meditation, inquiry, opening the heart.

He didn't want to hear it. If I tried having conversations with him about anything except the stock market or business, he'd say, "You're crazy," and look at me as if I were talking about things that only "those nutcakes in California" care about.

My friend Annie once told me that not everyone wants life to be a mountaintop experience. She said that we all get our emotional half acres to tend while we are alive. Some people grow potatoes, and some grow roses, but it's not our business what someone else does with their half acre.

After the first week, I had to accept the fact that my desire to share my life with my father was for me, not for him. I also had to recognize that not only was my father not going to be transformed by his illness, it was quite possible that he would never admit he had one.

The nurses' aides stop talking to each other as soon as Pepper walks into a room. The doctors ask to speak with my brother and me, not her. If she calls their offices, the receptionists complain to us afterward about her hostility. "She's so difficult," they say. "How does he live with her?"

Pepper snaps when you ask her anything, withdraws at the slightest hint of disagreement, gets offended easily. Last week she complained that she and my father were spending all their money to get him well, and when he got well, they would be poor. My brother, Howard, pointed out that since they have well over three million dollars, and they've only spent fifty thousand so far, her chances of being poverty-stricken are minimal.

Talking to Pepper is like walking on barbed

wire; it is painful each time, and therefore best to head the other way. Besides feeling murderous rage there is nothing to be done. My mother, though she is now remarried, takes my father to his treatments when I am in California. Once she had to leave halfway through a chemotherapy treatment, so she called Pepper at work and insisted that she come to the doctor's office. But as a rule, we know not to ask Pepper for help.

It's almost impossible not to think of my father as the victim and Pepper as Cruella De Vil, except that this is exactly how I thought of him and my mother for thirty years, which makes me the teeniest bit suspicious that there is a pattern here.

On the day of my father's third chemotherapy treatment, I beg a stranger to carry my father up four flights of steps. The elevator in the doctor's building is broken, there is no chair in the lobby, and my dad is cold and worn out and needs to lie down. I find this man walking in the building and tell him we need help and could he please carry my father up the stairs. He says no, then looks at my crumpled face and says, "Your father needs chemotherapy, I will do anything to help you." I want to cover his face with kisses. Right then I know that if someone I don't know is willing to carry someone he doesn't know up a hundred steps, love isn't just what you feel with friends and family. I've gotten it backward: love doesn't exist in physical form; form exists in love.

The next morning, a policeman starts giving me a ticket outside of Starbucks because I am double-parked in the sleet and the snow. I begin to cry. I tell him I just spent the night with my father, who is dying. After he put his summons pad away, he tells me about his twenty-eight-year-old daughter who died of ovarian cancer and left him with her three-year-old daughter. He also tells me that his twenty-one-year-old daughter has brain cancer. He begins to tear up when he says, "Sometimes you wonder if there really is a God."

Why didn't I try harder? Why? I woke up last night sobbing. Why didn't I visit him more in New York these past five years? Why didn't I go to my second cousin's wedding in June, where I could have danced with my father one last time? Why didn't I see he was depressed and do something about it? Why didn't I notice that he looked sick when I saw him in October, and insist he go to a doctor? I feel doomed to a lifetime of regret.

Jeanne says I have a choice: I can lean into the pain and allow myself to feel that the Empire State Building has landed on my heart. Or I can make myself insane because everything I am regretting has already happened, and it is too late to change the outcome. For an intensity queen like me, this is a tough choice.

Despite three chemotherapy treatments, the cancer is now in my father's stomach, adrenals, kidneys, and bones. As I sit in the Silver Moon Diner eating a piece of cheesecake, the taste of it shoots into my mouth like a star and I am suddenly grateful that cheesecake was invented.

Years ago, when I was struggling with emotional eating, I would have used food to get through this time. Each day would have been one long meal, punctuated by moments of being with my father.

Before, although I would have inhaled an entire cheesecake at one sitting, I wouldn't have tasted it. But this time I get to experience the pain of his dying *and* the taste of cheesecake. The cheesecake part is a good thing.

For ten years my father has been telling us that his estate is going to be split 50-25-25, with fifty percent going to Pepper, and the rest to us. Howard wants to secure the portfolio value; he says my dad invested in risky, high-tech stocks, which is how he went from being bankrupt ten years ago to having money, but since he is not capable of making quick decisions now, the money should be placed in blue chips, or at least in safer investments. Also, my father has not told us where his will is, what other investments he owns, or how much money his antique watch collection is worth.

I trust that my father will do what he's promised, has done it already. Howard wants to see papers. He wants to know that things are in order. He wants my dad to sign a power of attorney.

We sit down with my dad in his darkened living room and Howard tells him what he wants to talk about. My father says, "I beg of you. I beg of you. Please stop. Don't make me do this."

"All we want is to know where things are, Dad," I say. "And for you to talk to us about what you have already put in place."

"I beg of you. I beg of you. STOP." His voice sounds raspy, like Don Corleone in the garden at the end of *The Godfather*.

Confused by this response, my brother and I walk into the kitchen to talk things over. "This is so odd," I say. "What do you think is going on?" He shrugs his shoulders.

"Maybe he didn't do what he said he was going to do," Howard says. "Maybe he doesn't want us to know the truth."

"Impossible," I say.

When we walk back into the living room, my father says to us, "I would rather be put into an old-age home than deal with tension between you and Pepper."

"Okay, Dad," I say. "We'll drop it."

When Howard talks to me about wills, money, what's going to happen after our father dies, I feel like a woman in *Fiddler on the Roof* who circles in as someone is dying and sees dollar signs on the tips of Death's wings. I feel like a vulture. There will be time enough after he dies, I keep saying; how bad could it be?

The next day my father has a photocopied page of his will on the desk when I walk into the room. "See?" he tells me, and points to a paragraph that states that half the estate goes to Pepper and the other half to Howard and me. "It's all taken care of."

A few days later, he tells me that Pepper threatened to leave him and he doesn't know what to do. (He doesn't defend her anymore. He calls her "my dear sainted wife" with the bitterness of a man who's been duped.) Could he come live with me in California, he asks. I don't know what to say — it would be physically impossible for him to make a cross-country move — so I say, "If the doctor says you can, you can," and then remind him that he is not dependent on Pepper for help — it is *his* money the two of them are living on, *his* money paying for nurses' aides, *his* money that bought the apartment.

"But if she leaves, at least I can come and live with you," he says.

My mother is convinced that we are going to end up with none of our father's money, and Pepper will move to Arizona and live at Canyon Ranch for the rest of her life, taking mud baths with our inheritance. She thinks I should ask my father for anything I want NOW. She also thinks I should come into his apartment every day with a large shopping bag and begin to remove silver candlesticks, bowls, bottles. "If I were still married to him, they'd be yours anyway," she says.

One day, she arrives at the apartment with a shopping bag big enough to hold an oven, and eyes the room. "I'll do it if you won't," she says. "I won't stand back and let you end up with nothing."

I am glad to have a mother who is on my side, but I tell her there is nothing to worry about. She tells me that I have stars in my eyes.

Now, three months after my father's diagnosis, his doctor is pulling me aside and telling me that my father's calcium levels are so high that if he doesn't go into the hospital immediately, he could lapse into a coma and die within ten days.

I ask the doctor to meet with me in the next room, ask what she recommends.

"I am a chemotherapist. Chemotherapists do chemotherapy."

"But what would you do if this were your father?" I ask. "It's clear that the chemo is not working."

"I would take him home," she says with eyes cast down. "I would let him die peacefully."

I stare at the bookshelves, the diplomas, the pictures of children in navy blue striped shirts on the walls. I notice a pink pen on the desk, three spiral-shaped neon orange paperclips, a stack of papers with left-slanted scrawl.

"Give me a few minutes," I say. I have no idea what will occur to me in a few minutes, since my mind seems to have dissolved, but at least it sounds like there is someone home in my body, considering the situation.

She nods, calls the hospital, asks if they have any beds and if they can admit him this afternoon.

"Can't he get the calcium-lowering treatment here, so that he doesn't have to go into the hospital again?" I manage to ask. Saving his life is something I know how to talk about. I am in familiar territory now.

"Nope," she says, "he can't. And if he is going to be admitted, then he might as well have another treatment." I never wanted him to have chemotherapy to begin with — I believed it would rob him of any strength he had left — but it wasn't my choice. The thought of disorienting him by plucking him from home and putting him back in the hospital seems cruel.

The doctor is quiet. She seems to recognize that it's not just a patient she's talking about, it's my father, and that I am face-to-face with the decision of extending his life or bringing him home to die.

I go back into the room with my father, kneel down next to him, take his bony hand in mine. Valdeen, his day nurse, has dressed him after the doctor's examination and he is wearing a navy blue Mets hat, a brown-and-gray argyle sweater, and tan corduroy pants. I say, "Dad, these treatments are not helping, and we are at a crossroads. You can go into the hospital for more treatment or go home, in which case, unless there is a miracle, you will probably die soon. What do you want to do?" As the words leave my mouth, I feel as if I am someone else, playacting with my make-believe father, reciting the lines to inform someone who has no real-life relation to me that he is dying. He shakes his head, says yes, plays with his watch, picks at his skin. As difficult as it is for me to speak the words, it seems harder for him to hear them. I will not get an answer from him.

I call Howard, his medical advocate. He says, "Bring him home."

As I make phone calls about my father's

imminent death, I finally understand the gift of death: it erases everything but itself. Death demands your full attention, which means that all the worrying, nattering, wandering, obsessing, planning, thinking, and fantasizing stops. You can't daydream about a lost love. You can't start thinking about the size of your thighs. You can't worry that you will spend all your money on earrings. You can't wonder why your lover left you or if the stock market will ever rebound.

Sitting at my father's desk, I am not thinking about how I am going to live through this, I am doing what needs to be done, one step, then the next, as if there is no tomorrow. No emotion, no thoughts of anything but being with him. After a lifetime of terror about his death, there is suddenly no fear, no worry, no anticipation. It is not a problem that my father is in the next room, dying. Only a mind with leftover energy can convert what is happening into the idea of a problem.

That same night, I drive into Manhattan to have dinner with a high school friend. On the way back to the Island, I stop by my father's apartment to kiss him good night. When I walk in the door, he is sitting in his blue-and-green plaid robe, in the wheelchair, watching television. As soon as he sees me, he smiles a ten-thousand-megawatt smile.

"Hi, sweetheart," he says, "I am so glad to see you."

His gladness warms me like liquid sunflowers. Since his diagnosis, he has never said, Thank you, I'm happy you are here. I hug him, kiss him, sit down with him for fifteen minutes. He asks how my evening was, how I am feeling. I tell him how much I love him. He says, "I know that. It's you and me, baby, it's always been you and me."

Those are the last words he says to me.

The next day he is in a coma. I curl up on the bed next to him; his breathing is steady, hoarse, his eyes are closed. If I didn't know he was my father, I wouldn't recognize him.

I spend the morning with him, playing Frank Sinatra CDs, and telling him stories. If studies about people in comas are accurate, he can hear everything I say. I also read aloud from *The Kabbalah*, *Harry Potter*, and *The Tibetan Book of the Dead*.

I figure it's good to cover all the bases.

Also, since he can't talk back, I can read him anything I want.

When I call my friend Annie, she says, "It's not your father there. It hasn't been your father for a long time. You are there so that like Sister Prejean in *Dead Man Walking*, he can see you at the end. You are there so that the last thing he sees on earth is the face of love."

As I surround my father's bed with fat bursts of cream cymbidium orchids, the place in me that is not my mind takes over. I begin saying, "Nothing real ever dies, Dad, it's okay to let go. Don't be afraid. I love you, we all love you." Until now I've lived as a lonely lightbulb, where burning out is big trouble, but from this new place, death is a single twinkle-light going out in a string of thousands; the current is still vital, hot, alive.

My father's breathing is short and raspy; it sounds as if he is trying to suck the bottom of a milk shake through a straw. Pepper alternately holds his hand and writes travel tips to London on a yellow legal pad for her boss. Grace, the nurse, sits on the chair on the side of the bed. Nothing real ever dies, Dad.

At three in the morning, I decide to go back to the hotel; I know I need to sleep for a few hours and can't do it there. If you need to go before I get back, that's okay, Dad, do whatever you need. At six, Pepper calls me and tells me that his breathing has taken a turn for the worse, that I should return right away.

Hours pass. I read to him, talk to him, watch him. Valdeen changes him from his pajamas to a T-shirt that he bought on Orcas Island, during the week of my wedding. Except for the times we exercised together, I have never seen my father in a T-shirt, only sweaters, shirts, ties, suits. As I dampen his forehead with a washcloth, sprinkle water on his lips, I keep seeing the words "Having a Whale of a Time" on his shirt and I wonder what secret part of him chose, out of a million styles, this unabashedly corny shirt he wears, as we swim together in this whale of a time called dying.

Time loses meaning, becomes elastic, bendable, infinite. I sit beside his bed reading, singing along with Frank Sinatra, telling my father I love him. When I touch his arm, he pulls away. He seems to know he is going somewhere alone and needs to begin the journey now. In this way, hours pass.

At eight-thirty, Matt arrives from California. We replace Sinatra with Charlotte Church's *Voice of an Angel* CD. During the first song, my father lifts his head, opens his watery blue eyes, looks at me, turns and looks at Pepper, turns back, closes his eyes. A tear falls from his right eye. I am mesmerized by the tear, which takes a lifetime to travel down his cheek. I hear his sharp, raspy in breath, wait for the out breath. Nothing. There is sudden silence. Utter quiet. Pepper and Grace, the nurse, begin to cry. I am confused by their tears. Grace says, "This is it, 8:51 p.m., he's dead."

Is this death? The moment between breath and no breath? It doesn't feel as if anything is different — nothing has changed except that he stopped breathing. There is no sadness, only a sense of being enveloped in an unnamable something, a black velvet mystery that Grace is calling death. Is my father still here? Has he died? I suddenly have no idea what dying means. I look at Matt. He smiles at me without smiling.

Because it seems like the next thing to do, I read from Stephen Levine's book *Who Dies?* The piece begins, "Listen now, for what is called death has arrived. Do not be faint in this most precious moment. Recognize the natural radiance of being that shines before you. Enter into it. Old fears. Old bindings. Let them go . . ."

The reading is ten minutes long. I repeat it three times.

It occurs to me to ask Pepper if she would like to be alone with him; she says yes. We leave the room, and I call my brother, my mother. I say, "Dad just died." Three words. Dad just died. I am making the call in the night that I always lived in terror of receiving. Dad just died.

Six hours later, after dozens of phone calls to family, after my father's body is taken away, my brother, Matt, and I are walking out of the elevator to the front door of the apartment building. I am giving Howard a description of the evening: the last breath, the opening of eyes, the tear. The fact that he died ten minutes after Matt's arrival.

Howard, who has always been able to make me laugh faster and louder than anyone on earth, turns to Matt and says, "Geez, Matt — he was doing FINE until you got here!" I begin to laugh quietly at first, but then a snort escapes and suddenly it all seems so surreal — my father's death, the snort, Howard's joke, standing on a blue paisley carpet in a high-rise apartment building at three a.m. — I begin laughing so hard I am gulping for air. Howard begins to giggle, and as he giggles — a sound my nervous system recognizes from years of making faces at each other behind our parents' backs — I snort with greater frequency. Now I am gasping, holding my sides, begging him to stop laughing so that I can breathe. Matt's eyes widen in amazement. I can tell that he's torn between surrendering to this unexpected mirth and holding the line for appropriate after-death behavior. He watches the two of us for thirty seconds and dissolves into laughter himself. After we wipe the tears from our eyes, the three of us walk into the soft black night.

Six

Cheryl Schwartz, Blanche's vet, moved to San Diego the month before my father got ill, but as we are getting our luggage in the San Francisco airport after his funeral, we see her unmistakable profile: surfboard under one arm, saxophone under the other. She is in town to give a keynote speech about her book *Four Paws, Five Directions*. I tell her about my father's death; she says that when her mother was dying, her cat Hollywood helped her through the grief. She says Blanche will know what to do and I should let him comfort me. She also tells me that what's good for Blanche is good for me; am I carrying Rescue Remedy?

At home, the sadness is ever-present. In my flair for drama, I sometimes convince myself that my sadness is about wanting my father to still be alive. I assemble an altar on the living room end table with a huge picture of my father and me, candles, sympathy cards, flowers. I look at old pictures, resent every old person I see, wonder why they are alive and my father's dead. But when I am quiet, there is a deep recognition that he wasn't enjoying life on earth, that he'd been gone for a long time, that the timing was perfect. That death and dying are not about me or what I want.

At these times, I see that my sadness is about the direct understanding that things, people, cats, everything I love is temporary. The Buddha wasn't kidding. Einstein knew it, too. In a letter he wrote to a child who'd written to him saying that she thought he was dead, Einstein replied: "I have to apologize to you that I am still among the living. There will be a remedy for this, however."

After my father's death, I understand — not just with my mind, but with my cells — that the consequence of living is dying, that if he couldn't get out of it, no one can. No matter how much I meditate, pray, do weight-bearing exercises, eat organic food, wear ionizers on airplanes, panic about dying, sooner or later — by illness, tragedy, natural causes — it is all going to end. Even the longest life is a brief run.

The upside of this recognition is that I can't go back to the way I was before. There is no point in trying to get off the roller coaster and figure out how to work the controls, since there *is* no off and there are no controls. I could spend the rest of the ride throwing up or I could stand up, wave my arms, and have some fun. Either way, I will end up at the same place.

Blanche, of course, has known this all along. He doesn't seem to be worried about the fact that he is 105 years old and should restrict what he eats to keep his kidneys hydrated. He still pounces on dried salmon, bats mice around, surfs on light.

He follows me everywhere, and when I cry about my father's death and Matt is not there to hold me, Blanche curls himself next to my head, stretches out his paws to touch my face lightly, and purrs. To him, big gulping snotty tears are just another activity. After I finish crying, Blanche looks at me as if to say, "So what's next on the agenda? Pizza and yum-yum's? Sunroom nap? Motorized mice? Why don't we take a nap, then lounge around for the rest of the day?"

In the following months I notice again and again that Blanche doesn't act as if he needs to accomplish or succeed at anything to justify his life. Being present is enough for him, whereas I believe I have to earn love, rest, play, value. I still believe that suffering is noble, and if I work myself into the ground, I can take a nap. If I get sick, I can rest. If I succeed, I can feel valuable. I have to constantly *prove* I am worth taking up space, as if the prosecution has evidence to the contrary.

Blanche is my constant reminder that living on earth is not about deserving but allowing. His attitude is: Why insist on breaking your back to get to heaven when you're already here? In the months of grieving following my father's death, I spend entire days with Blanche, napping, eating (tuna fish for me, tuna fish juice for him), taking baths (during which he sits on the mat and after which he jumps into the empty tub and drinks from the faucet), wandering around the backyard, cutting flowers, and sitting on the bench in the garden. On these days I get the tiniest glimpse that Blanche knows the truth: I spend my life wanting things I already have — work, love, and chunks of seventy-seven percent bittersweet chocolate.

Two months after my father's death, I am on my way out the door, coat on, car keys in hand when the phone rings. Howard says, "I have some bad news. Dad lied to us about his will. He left almost everything to Pepper."

I don't believe him. It's as if a 747 is roaring over my head, losing gas, wings, engines, but lacking a runway. There is no space for this information to land.

Finally, I say, "This can't be true. I don't believe it. I'm sure it's a mistake."

"It's not a mistake," Howard says. "I've been over this with the lawyer very, very carefully."

"Dad must not have known what he was doing," I say.

"He knew. He was a *lawyer*, for God sakes."

Apparently, almost all his assets are either part of his pension, which means that Pepper, as his wife, gets the money, or designated to her as his sole beneficiary. He also did not specify anything about his personal possessions, which means that Pepper is entitled to all of them. She changed the locks to their apartment within twenty-four hours of his death and made it clear that except for the assorted items she knew my father had promised us — a wall sconce, some photographs, a painting, a magazine rack — the rest of his things, including the brass desk set my mother's mother had given him when my parents married — are nonnegotiably hers. We can have any of his clothes we want, but the rest is to remain in her possession.

I can't put them together: my father the mensch, the brilliant businessman, who loved me best, and my father the liar, the wimp who left his money to a woman he didn't seem to love, a woman he knew in those last few months — when it was not too late to change his will — had virtually abandoned him. To the woman who would not take him to the doctor he left three million dollars, and to his children and grandchildren he did not leave a cent (except for an insurance policy that Howard hounded him to split 50-25-25). Though I never thought of his money as mine, a lifetime of having a father who would stand in front of an oncoming train to protect me comes crashing up against reality. My father, who would never do anything to hurt me, who always told the truth, who knew best in any situation, was not, it seems, my dream of him.

For weeks after the phone call, I am mute and numb. I feel as if I've just found out he had a second life, with another family, in another city. Every now and then a flash of anger streaks through like a comet and I ride the blast for a few days, making Pepper my target. I walk around in a huff of blame: Pepper made him do it. Pepper knew this all along, that's why she never helped. Pepper stood back and watched us poor slobs turning our lives inside out for someone she knew had already betrayed us. My mother was right. I should have let her steal candlesticks, bowls, paintings, perfume bottles, silver.

I hate Pepper more than I've hated anyone, ever. Then I hate myself for being such a fool. For all the butterfly kisses, for all the excuses I made for him, for all the times I said, I love you, Daddy, you are my hero, Daddy, you saved my life, Daddy. For flying back and forth every two weeks to visit the schmuck who never even said thank you or good-bye. For not insisting that he deal with my brother and me *now*. For not shouting, DAD, YOU'RE DYING, STOP PRETENDING YOU'RE GOING TO GET BETTER,

AND DEAL WITH IT.

After that I lapse into making a list of all the times he copped out, didn't fight, didn't stand up for what he loved. I tell myself he couldn't help it, it wasn't personal, he was incapable of courage, and his mishandling of his estate was no different, why did I think it would be?

It isn't the money I crave; money is only the transport, the red caboose on which my father's love hitched a ride. I don't need a million bucks (although I wouldn't turn it down); I need to know I am *worth* a million bucks. I need his love to be so grand it illuminates me, fills me with absolute, irrevocable permission to take up space, be alive.

Days pass. I walk around in a woozy haze of sadness, betrayal, shock.

I don't want to examine my relationship with my father. I want to have had a living father who loved me and a dead father who was my hero. I want to be someone who was adored. Nothing short of Ward Cleaver will do. Given a choice, I'd rather be fat with face hair than see who he really was, and given that most women would rather be blind and get run over by a truck than gain ten pounds, that's saying something.

The tiny hitch in my lifelong denial is that by not leaning into the simple facts — he was not the person I believed he was — I never grow up. I continue to see myself as needing to be saved. And if my savior happens to have a few minor flaws — like being a liar and a wimp — well, hey, no one's perfect.

I can only recall fragments, chips of color: I am three years old. I am in the shower with him, watching the quilts of light stream in from the round cut-glass window. I am asking why I can't walk around with shampoo in my hair. He tells me the shampoo will attract bees.

I am eight years old. He is playing with me and his silver Jewish star swings around his neck like a metronome, as he walks into my room.

I remember his socks. Ribbed, thin black socks.

I am ten years old, I am walking into his office on 25 West 43rd Street in Manhattan. The doors have gold writing on thick wavy opaque glass, like the kind you see in Humphrey Bogart movies. He has a fish-lipped secretary named Riva with a low, growly voice who smells damp, like the inside of a cave. Everyone wants to see me, meet the president's daughter. They are frightened of him, cowed by him — Yes, Mr. Roth; Okay, Mr. Roth. He pulls phones out of walls, smashes his fists on desks.

I know not to say no or get angry at my

father. I've seen the consequences of his fury: he stops talking to my brother and my mother for days, sometimes weeks. Witnessing his stony silence is so unbearable that I vow to do anything he wants. I learn to turn my feelings off. I learn to pretend.

I pretend I am alive because he gives me permission to be.

I pretend I don't care when he doesn't talk to me or show any interest in what I think or feel or want.

I pretend it doesn't matter that he is never home. I pretend I don't notice that he ignores my brother.

I pretend I am not ashamed he can't remember any of my friends' names, not even to ask where they live.

I pretend he is the way he is because he works hard and does important things all day. I spend so much time pretending that I see what I don't see, that I like what I hate, that I no longer know the truth.

In her book, *Passionate Presence*, Catherine Ingram writes that her six-year-old friend asked, "Pretend you are in a room surrounded by tigers — what would you do?" She thought of every possibility — running away, scaring the tigers, calling for help, until finally she gave up and said, "Gee, I don't know — what would you do?" And he said, "I'd stop pretending!"

105

As children, there is nothing to do but pretend. We make up stories about why things are the way they are. If our mothers are depressed, if our fathers are drunks, we are still dependent on them for love and for survival. When the love is not attuned, when it is abusive, when it is scarce or intermittent, we often blame ourselves. We can't blame our parents because we need them too much; blaming them means rejecting them and rejecting them means dying. So we decide we are too needy, too intense, too fat. We were born defective. These things aren't true, but we believe they are. They get wired into our brains and hearts at the same time our nervous systems develop, and unless we question them head-on, we believe them for the rest of our lives.

My father and I had a secret contract: if I didn't say no to his requests, if I made straight A's and believed I was dumb, if I crowed over him and pecked at myself, if I was loyal, if I was sexy, if I thought of him as the source of my brains, my power, my love — my life force — I got Patti Playpals and fuzzy purple sweaters and a big weekly allowance. As I got older, I got a new Oldsmobile Cutlass and diamond earrings. There was nothing I couldn't have if I played my father's child mistress pussycat. For a pound of my soul, I got money, things, power.

All the women in my family — my grandmother, aunts, cousins, and mother — came to me when they wanted something from my father. They were positive if anyone could "get the goods out of him," it was me.

If I couldn't have my mother, at least I could have what she wanted.

Talk about winning the Oedipal battle.

My teacher Jeanne says it was "a choiceless choice." What else could a child do? she asks. She reminds me, however, that choiceless as it was, it was I who made the decision. She insists that no part of myself was forever lost because it was fundamentally mine, and while it may appear for decades that he took something precious, or that I gave myself away, she says that a human being cannot ever give away their essence. It is yours, she says, like your eyes are yours, like your skin. She says it wasn't the truth of my relationship with my father I threw to the bottom of the ocean, it was what was real and true in myself, and that until I reconnect with *that* I will always feel as if something is missing, no matter how much I am showered with love, adoration, affection.

That's the difference between a spiritual teacher and a therapist. A therapist sees the problem — you had a father who treated you like an object — and tries to help you heal the wound that came from it. A spiritual teacher sees the problem — a father who treated you like an object — and believes the wound is not in what he did, but in what you did to survive the situation. The wound is that you left yourself. The wound is that you forgot who you were/are in the desperate attempt to be loved. With a therapist, you have direct experiences of being your best self — loved, strong, worthy, whole. With a spiritual teacher, you have direct experiences of being the space in which that best self manifests. With a therapist, you learn that you are loved. With a spiritual teacher, you learn that you are love itself.

But, wait. This is why the word "drama" was invented.

Fuck being spiritual. Fuck being love. Fuck having to do this myself.

Sometimes a girl has to kick and cry and scream and make the most of her situation. My father was a narcissist, and a profound disappointment. Not only did he turn around and die without saying thank you or good-bye, but he left three million dollars to Cruella De Vil.

I could ride this one for a long time. I could lose myself in the waves of victimhood for months, years. There isn't one person I know who isn't on my side now. Finally. I get to be the good one, the innocent one, the wronged one. I get to be Melanie instead of

Scarlett O'Hara. For fifteen minutes or two years, I get to be absolved of all that I know is dishonest in me and feel only pure, only kind, only right. And after a lifetime of feeling damaged, this is quite a switch. It would be so refreshing to put all the blame out there. To say I am miserable because of *them*. The line between the genuine feelings and turning them into a Macy's Day Parade is barely perceptible. No one could blame a grieving girl like me for missing it.

Seven

After hearing stories about my tap-dancing, show-tune-singing daddy, when Matt met him for the first time my father sat hunched in a red leather booth of the Silver Moon Diner. Every now and then a couple of monosyllabic words would emerge from his mouth and he'd say, "So." "Matt." "What do you do?" after which he'd stop talking for the time it took us to eat three half-sour pickles and two pieces of pumpernickel raisin bread. I was busy being my magpie self — chattering to fill the empty spaces — and trying to pretend this was an ordinary conversation.

Later that evening, Matt, in an uncharacteristic moment, said, "You know, sweetie, your dad was nothing like I thought he would be."

Silence.

The description of the father I'd grown up with and the father he'd just met were so wildly divergent that at this point in our relationship — we'd only known each other for six months — I am guessing Matt felt it was necessary to check out his new lover's connection with reality. Because the next thing he said was, "The man I just met was a hair

110

short of being catatonic."

"He used to be the way I described him, he really did. It's just that he's lost his company for the second time, gone bankrupt, and retired, so his life is empty."

"It's more than that. I'm not a psychologist, but there was no one home there."

When my mother used to complain about my father, I locked up my mind, closed down my body — I won't let you tear him down; no, you can't take him from me — so that nothing she said could get through. When he went to work the day of my high school sweetheart's funeral, I excused him immediately and continued adoring him. The first time he heard me speak publicly, he told me that I had so much charisma, I reminded him of Hitler. I thought it was an odd thing to say, especially since he'd lost half his family at Auschwitz, but I dismissed it. He was my perfect daddy, and that was that.

That I kept my relationship with my father bullet-proof is not news. What *is* news is this: The reasons I gave myself for not being truthful about our relationship — he did not tolerate disloyalty; he would cut me off without a toss of his toupeed head — are no longer valid. Our unspoken contract that I remain loyal to him no matter what has expired.

He's dead. I can do anything I want.

A few months after the will debacle, Matt and I are lying in bed late at night. Blanche is taking up three-quarters of my pillow, jamming Matt's head and mine so close together that we are touching noses. I pick Blanche up and place him at the bottom of the bed; he picks himself up and walks to the pillow again. Because he weighs twenty pounds and is committed to taking the shortest route, Blanche's path involves trampling on my legs, stomach, chest, then using my nose and eyes as stepping stones. After three excruciating round trips, I surrender. Blanche spreads himself luxuriantly across my pillow — front and back legs extended to their full length — while Matt and I perch our heads on the section of his pillow that Blanche's paws have not yet usurped.

Our conversation for the past few months has been dominated by talk about my father. Did his rapidly deteriorating illness prevent him from changing his will when he discovered Pepper was not sainted? Did he want to leave us anything, or did he believe he had discharged his fatherly duties?

When Matt and I aren't talking about the will and the mystery of my father's intentions, I am trying to understand why everyone but me knew my father was part louse.

As Matt kisses my head, I can't help but

112

remember the many childhood nights I woke up crying and begged him to hold me because I was scared of being alone. My mother was always slightly resentful of these two a.m. intrusions, but my father would automatically lift the covers and let me crawl into bed beside him.

Matt breaks my reverie by asking me if I remember the first time I realized my father, the erstwhile business mogul, was spending his days in front of the television with his cat, Minnie, on his lap.

"Yep," I say, "I do."

I was sitting with Pepper and my father in the United Airlines terminal at JFK. I was picking at a fruit salad; my father was drinking black coffee with two Sweet'N Lows, and Pepper was eating a hamburger.

What are you doing with your days, Dad?

Nothing much.

What's nothing much?

I watch television.

Television? All day long?

Yes.

What about setting up a small one-person law practice? You've always wanted to do that.

I'll see.

What does that mean?

Leave it alone.

But Dad, you could be mentoring kids or teaching community college or reading to blind children or taking law courses at night.

He said, "I'll see." Now leave it alone.

But I couldn't leave it alone. I called my literary agent, who had mentored a child in New York. She gave me a number for my father to call. I called him and gave him the number. He said he would call. He never did. He said he would take a six-week computer class. He went one night and never went back. I told Matt that I would fly to New York and take my father around to volunteer offices for a few days, sit with him until he found something he loved to do. Matt said, "You have to let him live his own life."

I didn't go to New York. Not because I believed the business about letting my father live his own life, but because he wouldn't talk. After the hello kisses and the how are yous, he disappeared into a leaden silence for fifteen or twenty minutes at a time. He was never a big talker except about himself and the topics that interested him: real estate, the stock market, the Duke and Duchess of Windsor, antique watches, Ralph Lauren socks, and Paul Stuart suits. After he married Pepper, conversation became so scant that going out to dinner with them felt like drowning in an abyss.

Blanche stands up on the pillow, stretches his legs, and places one of his front paws on the back of my neck. As he inches toward us,

Matt and I are perilously close to falling off the bed. "You weren't close to your dad in the last fifteen years before he died. You didn't like spending time with him," Matt says.

"It's true," I say grudgingly.

Even after all that's transpired, I still cling to my glossy ideal of him.

During the first six months of life, every infant develops a defense mechanism called "splitting." Since there is no way to get up and walk out of the room — the most a tiny baby can do is turn her head away from something she doesn't like — and because mothers are imperfectly attuned to our needs, our hungers, our hurts, we develop the ability to think of our mothers as two entirely different people, the good mother and the bad mother. We create an impenetrable mental boundary around the bad mother so that our pain and disappointment cannot contaminate the love we receive from the good mother. In this way, our nascent nervous systems do not take on more frustration than they can handle.

As we get older, we sometimes revert back to splitting during fights with our friends, children, spouses, and parents. For instance, when Matt and I are having a fight — he's screaming at me, I am screaming at him, both of us are making quite a ruckus — I suddenly forget every good or kind thing about him. One moment he's my beloved husband, and the next moment, I hate his voice, his hair, his face. If I am on a roll, I convince myself that marrying him was a huge mistake and as soon as I can call a divorce lawyer, I'm leaving.

Eventually, I calm myself down or Matt makes a joke. One time he cracked a raw egg over his head in the middle of a fight, which is when I discovered that being angry and watching an egg yolk drip down someone's face are mutually exclusive.

My insistence on seeing only the good in my father was an extreme example of splitting. For years, I believed that if I allowed myself to see anything unkind about him, I would lose him altogether and be back in the clutches of my mother, whom I'd placed in the all-bad box.

They were a team inside my psyche, my mother and father. They were counterparts, two sides of a coin. I needed him to be the perfect daddy who adored me so that I could survive the cold mother who hated me. But here's a million-dollar question: If my father wasn't all good, how could my mother be all bad?

Because despite years of therapy, despite years of healing between us, despite the fact that it is my mother who keeps showing up, who picked out Blanche, who has never, not once, forgotten my birthday, who would never dream of writing me out of her will, my default position is to reject us both.

At the end of a session with Jeanne during which we work on the tenacity of my beliefs about my mother and me, she reminds me of the studies of abused children who cling to their abusers rather than trust someone who is kind and loving. She tells me that there is something inside the child who has been well loved that knows how to relax, whereas children who have been neglected or hated or abused (or believe they have) hold on tight for so long, they don't know how to let go.

I see this pattern when people get within five pounds of their natural weight and start bingeing. It's as if they are hard-wired for disappointment and failure; nothing in their system recognizes success, and so when it comes, they sabotage it immediately and return to the familiar ways they know themselves: as being separated from what they most want, as yearning for what they don't have, and as feeling hopeless about ever getting it. Although it seems as if we should feel ecstatic when we finally get what we want, we often feel an undercurrent of fear, loneliness, and anxiety at the prospect of losing our old, familiar selves.

Jeanne says that eventually we let go of

what is hurting us because holding on becomes too painful. It's like gripping the handle of a cast-iron frying pan that's been on the stove for three days. We suddenly realize that our hands are being seared and that dropping the pan would hurt less than holding on tight.

As I am walking out the door, Jeanne says, "Only people who are dead or enlightened are done with their mothers." I feel better instantly.

Eight

Four months after my father's death, Matt, Blanche, and I move to a house in the country, where there is plenty of space for Blanche to romp. Blanche's new sport is catching the multitudes of mice who tear across the floor every few hours, chewing off their heads, and leaving their torsos for us as gifts. Years ago, Cheryl told me there was too much calcium in mice for Blanche's system, but when I call her about the mice heads, she says it is good for him to have stimulation and that catching them will keep him spry.

The house comes equipped with a small black cat with white paws and a white diamond on his forehead. We see his face pressed up against the window just as we wake up every morning, while Blanche is splayed on the bed after another night's sleep on a football field–sized down comforter. Matt calls them the Prince and the Pauper.

When we walk outside, the Pauper runs over to our legs and purrs. Even when I don't stop, don't look at him, tell him he needs to find somewhere else to live, the Pauper keeps purring, keeps rubbing, and keeps following me. Like Blanche when I brought him home, the Pauper is unfazed by the fact that his affection is unwanted. He reminds me of what a teacher of mine once said, "Love gives away the store."

But just as I did that first week with Blanche, I refuse to be taken in by the Pauper's affection. As I walk past him, I tell him that this new house is Blanche's territory, although it seems I am the only one threatened by the Pauper's presence. Blanche ignores him and Matt thinks he is cute.

The Pauper is a waterfall of cheer, love, affection, and the more I seal myself off from him, the worse I feel. After a week of seeing the Pauper's face every day outside our window, I cave in. We give him a name — Binky — and buy him an igloo house that we carpet and furnish with a bed, catnip toys, and dishes with his name on them.

Blanche still looks right past Binky as if he is not here, but every morning I go out to see him. I pick him up and stroke his head. It seems my heart has two positions: walled off or wide open. Which brings me back to my mother. A few weeks ago, Matt suggested that I ask her to come for a visit during his next business trip. He said, "She knew your father best. If there is a puzzle here, she probably has some clues."

The last time my mother visited, we talked, watched *All My Children*, and went shopping. She calls it "shpatzeering" — a Yiddish word for desultory roaming, eating, and talking. During one of our retail jaunts, she bought me an elbow-length pair of gold lamé gloves. Although my mother is usually an excellent shopper, the gloves were not our finest moment. The two times I wore them, I felt like Liberace when he said, "I did not come here to be unnoticed."

I call my mother and ask if she wants to come for a long weekend. I tell her I want her to see our new house, and go shpatzeering. We set a date two months away.

After a few weeks of being in our new house, Matt says he wants a dog.

"A dog?" I repeat.

"A BIG dog," he says gleefully.

I decide it is best to act like my half-deaf grandmother. When she didn't want to engage in a conversation she turned off her hearing aids and pretended she couldn't hear you. If I ignore the dog discussion, perhaps it will go away.

It doesn't. So I pull out the stops.

I tell him I don't want a big, slimy, licky, hairy, lumbering, smelly creature running around my house.

I tell him Blanche is too old to have siblings.

I tell him I am never going to be able to love another animal the way I love Blanche — there isn't enough room in my newly discovered heart.

I tell him Blanche will be so jealous it might kill him, and if it kills Blanche, it kills me.

Matt reminds me he's always been a dog person, and Blanche is an exception. (I agree.) He reminds me he doesn't like cats, but Blanche is so unique, so funny, so engaging, and so full of personality that anyone would be crazy not to love him. (I can't argue with that.) He says Blanche is more like a dog than a cat. (I believe that is taking things too far.) And then he launches into the Madeline story.

Madeline was the black Labrador that he and his college girlfriend, Sue, picked from a basket of puppies outside of Safeway. Every day on his way to classes, Matt would drop Madeline off at the corner of University and Telegraph and she would wander around the shops, visiting friends. She met Matt again at a designated time and agreed-upon corner after his day at school. When he and Susan broke up and moved to different coasts, they shared custody of Madeline for five years by shipping her back and forth every six months, until a pair of dog-sitters Susan hired wouldn't give Madeline back. Those evil dognappers even changed the dog's name and moved houses so that Matt and Susan could never find them. But find them they did, and when Matt showed up at their house, Madeline (a.k.a. Fluffy) ran to him immediately, despite the dog-sitters' protestations that this was a completely different dog.

On the wall in Matt's office an eight-by-ten framed photograph commemorates Matt with a huge afro, and Madeline with a bandanna tied around her neck, touching noses; every few months, he tells me about the days when despite being told she couldn't sleep on the water bed, that was where she parked herself. However, she was smart enough that as soon as she heard Matt coming in the door, she'd jump down and pretend she was sleeping on her own bed. But then Matt would notice that the bed was rocking on its frame, and Madeline would cover her ears with her paws so that she couldn't hear the unkind words he'd say next. I imagine it was something like, "Oh Madeline, you naughty pooch. Come here and touch noses with me."

It is becoming apparent that I'd married a dog person. A BIG-dog person. Nevertheless Matt had willingly entered into marriage with a woman whose first loyalty is to her preexisting offspring. The thought of bringing a puppy into our harmonious family is overwhelming; versatility is not one of my strong points. I've been eating the same breakfast every day for five years, and find great comfort in constancy.

On the other hand, when I look around at the rest of the world, I see throngs of people with two dogs, a couple of cats, and a gaggle of *children* — little humans, who, I am certain, need more attention than a puppy. Years before, I'd gone into an illness-induced early menopause and been told I couldn't have children. Since Matt never yearned for children, we decided to focus on our work, Blanche, and each other rather than try to adopt. Our life settled into a sweetly balanced equilibrium. Until now.

It humbles me to understand that I am in the same situation (except for differences in gender and species) as the woman who wrote: "My husband said it was him or the cat. I miss him sometimes."

Matt starts the puppy search. He calls friends, searches the Internet, goes to the dog rescue give-away on Sundays outside of George's Dog Accessories store on Fourth Street. I tell Matt that if I have to have a dog (and I do), I want one that doesn't shed — which leaves poodles, bichon frises, wheaton terriers, and Labradoodles, none of which fit Matt's image of a *man's* dog.

My friend Catherine Bacon has an eleven-year-old standard poodle named Lulu. She'd carried Lulu around in a snuggly for the first few months of her puppyhood, when she had a respiratory infection and was too weak to walk. Catherine is an art-to-wear clothes designer — Ellen Burstyn wore her clothes for the 2001 Academy Awards — who designed "The Lulu Coat," a puckered velvet swing coat designed to look like Lulu's hair, which I'd bought and worn for ten years. She also designed The Lulu Sweater, Lulu pants, and Lulu vests. Her bathroom is filled with poodle pins, poodle pictures, poodle paintings, poodle bowls. Compared to Lulu, Blanche is underprivileged.

At my urging, Matt agrees to meet Lulu.

Lulu is calm and obedient, with hair that resembles Matt's — curly, frizzy, and round. Her tail is in dreadlocks and her face has a bushy beard and Groucho Marx eyebrows. I watch Lulu with all seven of Catherine's cats, trying to assess the danger of bringing a dog into Blanche's life. She doesn't chase them, frighten them, or eat their food. In fact, Snaggle Tooth nests on Lulu's head the same way Blanche does on Matt's at night, paws hanging over his ears. (Once Matt dreamed that he was being buried alive, but when he opened his eyes, he realized Blanche had fallen asleep on his face.)

After we met Lulu, each of us was satisfied that our requirements for a dog were fulfilled: I was assured that we could train a non-shedding dog to be gentle with Blanche, and Matt was assured he could have a BIG dog without pom-poms. Within a week, we'd called the owner of an apricot poodle named Lily whom Catherine met at a park, discovered Lily was pregnant, and put ourselves on the list of potential parents. When a litter of twelve puppies was born, Blanche's sister Celeste was among them.

In the ten weeks between Celeste's birth and her arrival at our home, my mother comes to visit. During her second night here, we are eating butterscotch pudding at the Larkcreek Inn. The conversation turns to her relationship with my father. I ask her to tell me what she loved about him. I tell her I want to be realistic about him and to stop seeing her as the bad guy.

"You mean I'm going to get a reprieve?" she asks.

"Not exactly," I answer.

"Just tell me one thing. Would your father roll over in his grave if he knew about this?"

"Probably," I say.

"Well, I've gotten the hard knocks for a while now. I suppose it's his turn." She puts her spoon down, tilts her wineglass toward the ceiling, and says, "Here's to you, Bernie."

Then she leans toward me and says, "He was funny and charming and so very quick. Had a mind like the wind, agile and fast, and style, such style. It was like cutting off my right arm when we got divorced, but I had to. I had to. We were like gasoline and fire, blazing with cruelty, alcohol, insane fights, slammed doors. I begged him to go to therapy with me, but he refused. I'd met Dick, and saw that I could love and be loved, and I wanted that. In my heart of hearts, I believe Bernie wanted that, too, but he couldn't act on it. He just couldn't. If you ask me, that was the great tragedy of his life: He only let himself want what he knew he could get. He couldn't take risks. He couldn't be vulnerable, and without those kind of exchanges, love is just a sentimental wish.

"I almost went back to him right after Dick and I married. Bernie wrote me letters and told me how much he loved me and how much he'd changed, and God knows, I still loved him. I've always loved him. Lack of loving was never the problem. The problem was that you had to infer Bernie's love. He never showed me he loved me while we were together."

"But why do you think he showed that he loved me?" I ask.

"You were different. He could mold you. You didn't challenge him. He adored you from the second he laid eyes on you."

"So, Mom, were you jealous of my relationship with him?"

"No, never jealous. I was glad you had a

father who loved you that way. But I never underestimated his fury. One reason I didn't divorce him before you were nineteen is that I was afraid he would have taken it out on you and your brother. I knew he wasn't past cutting you off without a dime."

She'd told me this before, but I'd never believed her. Just like I had never believed he would lie to me about his will. I'd discarded her nasty stories about my father as desperate attempts to destroy my loyalty to him. Like the time he threw a diamond ring in her face. Like the time he asked her to get a coat hanger abortion in Spanish Harlem.

"After what happened with the will," I say, "I actually believe he might have done that."

"Not just the will," she says. "Remember when he stopped talking to you because you mentioned Dick's name?"

"I tried to forget about that."

After my parent's divorce, my father told me that I was not allowed to mention Dick's name *ever*. He told me that Dick was responsible for breaking up their marriage (even though my father had been having affairs for twenty years, even though he refused to go to therapy, even though Dick offered to quit the scene if my father wanted to try again with my mother). Thirteen years after the divorce, I accidentally mentioned Dick's name and my father stopped talking to me for four months. I didn't allow myself to think about what it felt like to have a father who would stop talking to me because of his hatred for another human being. Or because of anything at all.

Sitting there with my mother, I realize for the first time that it's possible my father might have disappeared if she had divorced him when my brother and I were children. Taking the high road was definitely not one of his virtues. He'd started his last company by illegally using his clients' mortgage money. If his venture had failed, dozens of people would have lost their homes. After a nasty divorce from my mother, it's possible my father would have convinced himself that he owed her — and us — nothing.

In the space of a ten-minute conversation, my mother has gone from someone who tried to take my father from me to someone who tried to keep my father with me.

I built my life — an eating disorder, a teaching career, six books, my self-image — and my relationship with my father — on having a mother who hated me.

When I stop insisting that my mother threw my heart against the wall and gloated when it shattered into pieces . . .

When I stop pretending that my father was the keeper of everything my mother lacked . . .

When I stop defining myself by what they

did and didn't do, and by my attempts to convince myself, my family, the world that I am not the child I saw through my parents' eyes . . .

I feel as if I've popped out of my own skin, as if I'm on the other side of the cage I've called home, floating in a universe without gravity.

My ideas about myself are my compass, my true north. I know who I am by my wounds and the friction of working against them. I know who I am by what I lack and how hard I try to make up for it.

Without my beliefs as protection, without insisting on seeing the world through my self-constructed lens, there is only what's here, at this instant. The pulsing sensations in my arms, my legs. The sharp coolness of air. If I am willing to stay with the sensations and not reassemble myself into a familiar configuration, I get to see things as they are — and always were, before I became a contortionist for love.

My mother and I spend five days together, during which she is a perfect grandmother to Blanche: she coos to him, feeds him baby food with her fingers, and carries him everywhere. I begin to realize that we've been good friends for a long time now. It slowly dawns on me that in continuing to shower my father with glitz, I've been depriving my-self of authentic gold — my relationship with my mother *now*.

Blanche and I are both misty-eyed when my mother leaves. I, because having finally found my mother, I am not ready for her to go, and Blanche, because his new means of rapid transportation has suddenly disap-peared.

Nine

When his new canine sister walks into the house, Blanche swats her, chases her, growls at her, then folds his paws underneath his body and glowers, as if to say, "I have no idea who had the crazy idea to bring you here, little missy, but I can assure you, it wasn't me." After a few swipes on her nose, Celeste understands that Blanche is not a chew toy. Occasionally when he isn't looking, she sidles up to his face and gives him a quick lick, but since that always results in him cuffing her ear, she soon realizes that the optimum distance from Blanche is across the room.

No matter how many times Catherine Bacon reassures me that Blanche will live through this invasion, I am certain he believes I've forsaken and betrayed him by bringing this dinosaur into our house (even as a puppy, Celeste is about three times the size of Blanche). For a few hours every day, I pick Blanche up and take him to my writing studio, where he can lounge on the chair and walk on the computer keyboard in peace. I remind him that though Matt wanted a dog, Blanche will always be my first child, my soul cat, my familiar.

We also try a technique we've read about in dog training manuals: associating an unpleasant activity with a treat. Every time Celeste walks in the room, I give Blanche a yum-yum. Though he polishes off twelve bags of yum-yums within a few weeks, he doesn't fall for my tricks. To Blanche, Celeste is a pariah, an intruder, a behemoth with bad breath.

Every year on her birthday, Catherine gives Lulu a party. She invites all ten of Lulu's children, plus any other standard poodles that she's met who don't bite or hog attention. When we receive the invitation for Lulu's twelfth-year celebration, Celeste is six months old and has never been to the groomer. As adult dogs, poodle hair turns wiry and curly, but as puppies it is stick-straight and very long. Annie says that with her huge red hair and spindly legs, Celeste looks like Tina Turner.

For the party, Celeste wears her rhinestone-studded collar and pink bows behind her ears. Bosco, Lola, Rhoda, and Jack are dressed similarly, though the boys have bow ties and not all the girls are wearing jewels. For a short while, it looks like poodle mayhem, with fourteen dogs running up and down the driveway, in and out of puddles, chasing cats, pouncing on frogs, smashing flowers. Then Catherine tells everyone it is

133

time to come inside and cut the cake.

The room is decorated in banners and streamers and happy birthday signs. Catherine has set the table with fourteen chairs, and instructed each of the dogs to hop up on a chair and wait for further instructions, which they all do. Then she explains to them that it is Lulu's day.

All heads turn to Lulu, who is wearing a gold tiara and listening to Catherine patiently. Celeste sits with her head cocked to one side, trying to figure out what all the people are doing standing up around a table of sitting dogs. Jack keeps trying to push Rhoda out of her seat, and Lily is so bored, she rests her chin on the table. Next, Catherine brings out the cake: a mountain of dog food iced with pink cottage cheese.

We sing "Happy Birthday." Jasper barks. Lulu waits. Celeste watches. Sam howls. Rhoda tries to push Zack out of her seat.

Catherine places the pieces on black paper plates, one hunk in front of each dog. Celeste bends her bow-bedecked head and chunks off the pink icing in a few slurps. Zack eats his piece in one huge bite and starts chomping on Rhoda's. Snaggle Tooth tears through the house. The commotion is too much for Sam, and as the demure birthday girl, Lulu, watches, he leaps onto the table and gobbles the rest of the cake before anyone can stop him.

As I watch this madness, I understand why Matt needs a dog. They frolic and leap around the house like he does. They are happy all the time like he is. He co-wrote one book called *Work Like Your Dog*, and another called *Dogs Don't Bite When a Growl Will Do*. Not only does he *look* like Celeste, he lives like her as well.

Of course he has to have a dog. Blanche and I are self-contained and finicky. Matt needs a slurper in the house.

When Blanche begins crying for hours every night and drinking from the bathtub faucet for ten minutes at a time, I begin looking for a vet who makes house calls. We had tried taking Blanche to an acupuncturist in San Rafael, but by the time we got to the bottom of our driveway, he'd thrown up, had diarrhea, and was crying piteously. We decided that the liabilities of traveling with him negated the benefits of treating him, so we stopped taking him in the car.

Dr. Gary Riden finds two tumors on Blanche's thyroid gland, and after reviewing his blood tests, tells us that unless we have them removed, Blanche's hyperactive thyroid will cause him to die of a heart attack within months or possibly weeks. He tells us about a nuclear medicine facility in Sacramento where cats can have their thyroid tumors zapped by radioactive isotopes; he says most cats do well with the procedure, and, after spending a week at the facility, are able to come home and live many more years.

Blanche hasn't been away from us since he had feline urinary syndrome when he was two years old; I can't imagine him living in a cage for a week. Who will give him slices of organic chicken every night? Who will smear turkey baby food on her fingers when he isn't eating enough out of the bowl? Who will make sweet potatoes for him when turkey and chicken aren't what he wants?

When I consult with Cheryl, she tells me that there isn't any choice but to give Blanche the operation. She says if I talk to him, he will understand that he needs to go away if he wants to feel better. She reminds me that twenty-pound cats with kidney disease rarely live longer than ten years. Given that he'd already eclipsed all predictions, it is possible he could live to twenty, twenty-one, twenty-two. "He's a miracle cat," she says. "You can't predict these things."

We arrive at the nuclear medicine facility with Blanche, accompanied by Celeste, twelve jars of baby food, two pounds of sliced chicken, cooked sweet potatoes, and dry food, plus Blanche's favorite blanket and the T-shirt of mine on which he takes afternoon naps.

Matt and I fall into our usual stances: he

thinks things will be fine, and I think Blanche will die immediately. I still can't imagine life without him. I delight in Celeste, but Blanche and I are knit together. He and I share the same unspoken language; we talk through walls, across the yard, in dreams. We *recognize* each other, and though he didn't have a say in the matter, I still feel chosen by him. Over and over, in a thousand different ways, Blanche's whole being — his eyes, his walk, his willingness to be picked up, to curl on my chest and jump in my arms — says, "It's you, I'm so glad it's you." Blanche is still my most tangible evidence that grace is possible.

A few weeks after Blanche returns home without his tumors, Gary tells us that his thyroid is normal, but his kidneys have now deteriorated and we need to hydrate him daily by giving him subcutaneous liquids from an IV bag. The procedure involves hefting Blanche onto the kitchen counter, placing a needle under his fur, and allowing five minutes for the solution to enter his body. He yowls, hisses, nips at our fingers, and then settles down. Afterward, he gets a treat.

Within a few weeks, the hydration procedure becomes so much a part of our daily routine — brushing our teeth, meditating, eating, giving Blanche liquids — that we don't think about it. When friends see us with IV bags, needles, and Blanche on the counter, they ask if he is okay. I always say yes. I know that he would have died without the operation, and I know that his kidneys will fail without the liquids, which makes him fine now. I've been told that some cats live for years on subcutaneous liquids and I assume Blanche will be among them.

One night, as we are falling asleep, Matt says, "You know, I think Blanche is going to be around for many more years."

I don't tell him to snap out of his optimism. I don't tell him to get with the dark side of life's program. I believe him.

Blanche is turning seventeen years old and I want to celebrate his life. Pumpkin and June, his mother and sister, have been dead for years. U.S. presidents have been sworn in, served, and disappeared; one was impeached. Babies born the same year as Blanche are applying for college. During his lifetime, computers became a household item, the Chicken Soup books multiplied like rabbits, Dr. Atkins made a second comeback, and Mr. Spock finally died. Blanche has lived longer than some humans, most dogs, hamsters, parakeets, and newts.

I figure that if dogs can have parties, cats can have them, too. Then Matt tells me that ten million people give their cats birthday parties every year, and that clinches it. We begin planning a party that is small, intimate, and unstintingly divine.

On the evening of the fête, Blanche's friends sweep into the house with gifts of organic catnip, toys with removable skirts and hats, and an orange-feathered fanning stick with which to do feline prostrations. Blanche's godfather presents the honoree with a crimson-and-gold king's crown that he'd spent the day cutting and pasting and duct-taping together from a person-sized coronet.

Next, the guests are assigned to draw cat eyes, whiskers, and noses on the person whose name they pick from a hat. When everyone is in full feline face, we give them a set of cat ears and tail to wear for the rest of the evening.

Though they'd viewed all three of Blanche's portraits before (maybe five or thirty times), we instruct our guests to see them with "beginner's mind" — a term that Zen master Shunryu Suzuki Roshi coined to describe the freshness of seeing something for the first time, without ideas, preconceived notions, or agendas. After the tour we scrunch on the couch and watch the video Matt made me as a birthday gift years before: a collage of images of Blanche walking, blinking, sitting, playing, running, drinking from the sink, being kissed, being hugged and thrown over our shoulders, all to the tune of "Dooh Wah Ditty."

Post-cinema, we toast Blanche's life with champagne and sushi while Blanche lies on his white mohair blanket next to the table and devours a bowl of tuna fish juice, dried salmon, and sweet potatoes. As the guests say good-bye, the women shower Blanche with lipstick-head kisses.

Three months later, Blanche starts wailing again at night, a piercing, heartbreaking cry. At first we think he is unhappy because his legs aren't strong enough to jump onto the bed directly from the floor, so we roll the faux-leopard tuffet next to the mattress and create a gradual stepping path of cushions to our heads. But in the middle of every night, he starts to cry while standing on the tuffet, and when I try to comfort him, he can't settle down.

Cheryl Schwartz prescribes new homeopathic remedies.

I hire an acupuncturist who makes house calls for cats and consult with a veterinarian in Lake Tahoe.

Nothing works.

After a few sleepless weeks, Matt suggests putting Blanche in the sunroom at night so that I can get some sleep. (Though he, Matt, sleeps through earthquakes and fires, I wake up if the refrigerator hums too loudly.) The thought of putting Blanche in the sunroom makes me miserable. I buy industrial-strength earplugs, a loud air filter, and two white-noise machines, but still he wakes me up three, four, five times a night.

Then Matt suggests we move Blanche's favorite toys, his white blanket, and dishes of food he likes into the sunroom with him; he says that since it's Blanche's favorite room and the place in which he spends hours each day, it won't be difficult for him to spend nights there. I try to believe him, but I know it isn't true — cats never like to be in a room with the doors closed, and Blanche is no exception. Still, I am ragged from not sleeping.

When Matt begins closing the doors of the sunroom each night with Blanche in it, I feel as if he is locking *me* in there. Sometimes I sleep through to morning, but most nights I wake up around two or three a.m. and check on Blanche. He is always alert, crying, waiting to be let out of the room. I drag a futon into the room so that I can sleep on the floor with him. Sometimes we both fall asleep. Then I wake up, close the doors behind me, and go back to bed, after which Blanche starts crying immediately.

I remember going to the hospital to get my tonsils out when I was five, watching my mother walk out of the room, and feeling helpless to stop her. I wanted to grab the hem of her skirt or latch on to her ankle; anything was better, even being dragged across the floor, than being left alone in the dark. When I leave Blanche in the sunroom at night, I feel as if I am abandoning myself over and over again. We follow this routine for three months and it is torturous every single night.

Ten

On three successive mornings, I notice Blanche hasn't touched his food, so I call Gary, the home vet. During a follow-up phone call after his visit, Gary tells us Blanche's creatinine (the toxin excreted by the kidneys) level is ten times higher than it should be. When he says, "I'm so sorry" three times in rapid succession, I feel as if I am encased in amber. Finally, I realize he is telling us Blanche is dying.

I hang up the phone, walk directly to Blanche, and beg him to eat. I notice the tears streaking down Matt's cheeks. I open packages of salmon flakes, convinced that if Blanche could eat, he could get stronger. I remember my father's face in the hospital when I asked him please to eat rice pudding. I swing from disbelief to acceptance, from begging Blanche not to die to being unspeakably grateful that he's stayed around longer than anyone imagined he would.

When Cheryl Schwartz receives a copy of the blood tests, she calls and says, "In true Blanche fashion, he has outdone himself. Though I usually don't take tremendous stock in numbers, these are staggering." She tells me that Blanche has been living on love interspersed with acupuncture treatments for years. She says Blanche has taken care of me for a lifetime, and now it is my turn to do the kindest thing for him. "With results like these," she says, "he has to be uncomfortable. Can you find it in your heart to allow him to go quickly, painlessly?"

It doesn't matter that I knew this was coming.

It doesn't matter that Blanche has already lived a long life.

The moment of hearing that someone or something you love is dying is new every time.

The same thing happened with my father: one day he and I were eating hot pretzels with mustard on 56th Street and two months later he was dying. Just like that.

It is Thanksgiving week. Matt's eighty-four-year-old mother is visiting. As she watches me stroking Blanche, carrying him from room to room, and trying to feed him by hand, I hear her whispering, "I hope someone does this with me when I am dying."

For three days Blanche sits quietly in front of the fireplace, staring into the ashes. At night we lift him onto the bed, where he sits on my pillow but doesn't sleep. I talk to Celeste's vet, who tells me he doesn't believe in euthanasia because it robs animals of facing their own deaths. Since I don't want to rob Blanche of anything, we decide to postpone making any decisions.

After six days of barely moving, Blanche begins to get restless and agitated. He tries to walk but his legs keep buckling. Then he falls and cries. We lift him into the litter box, after which he wants to get out. Once he is back on the floor, he wants to get back in the litter box or walk across the room. Since he can't walk, he falls on his chin. Nothing — not stroking, not petting, not singing, not carrying — calms him.

I am leaving to teach a five-day retreat that has been scheduled for a year. Cheryl Schwartz is traveling and can't be reached. Celeste's vet reassures us that falling and crying is part of the death process for Blanche and we should allow him to go through it, but it is difficult to understand how this is serving him. The grief of seeing him fall, the fact that I am leaving, and my conflicts about euthanasia are excruciating.

My friend Lily suggests that I ask my co-teachers to take over my sessions at the retreat and stay home with Blanche so that I can be with him at the end. But staying home, despite my feelings for Blanche, is not an option.

Ram Dass once said that whether we are sailing into the Aquarian Age or heading straight into Armageddon, the work is still the same: to quiet our minds and open our hearts and relieve the suffering around us. I love my cat, but I love the work Ram Dass refers to every bit as much.

Seventy-five people, many of whom I have been meeting with twice a year for four years, have committed to spend five intensive days to discover if there is something, anything, they can trust besides being thin, being rich, and hoping that what they love will live forever. Staying home now would speak louder than any teaching I could give.

Ten minutes before I leave, we rig a makeshift snuggly from a blanket whose ends we tie together around Matt's neck. As I walk out the door, Matt is sitting on the blue leather chair with a Blanche mound on his chest. Except for one paw that sticks straight out, Blanche is completely covered by the blanket. With his body molded to Matt's heart, he stops thrashing and crying for the first time in fourteen hours.

That night, a few hundred miles away, I realize I want to keep Blanche alive for me — so that I can see him one more time — not for him. Though Celeste's vet doesn't believe in euthanasia for animals, I realize *I* do.

After teaching the evening session, I call Matt, who has no doubt that putting Blanche down is the right thing to do. Then I call Jeanne and ask if she and her husband, Paul, could be with Matt the next day when Blanche dies.

The next day, between sessions at the retreat, I find a bird's feather, two rocks, and a candle for Blanche. Paul, a Zen Buddhist priest, instructs me on which objects are needed for an altar and what each signifies: the sky, the earth, and light. Though Zen is not my tradition, I am grateful for the structure of the ritual, since my mind seems to have washed away.

The vet is scheduled to arrive at my house, two hundred miles away, at two p.m. At ten minutes to two, my co-teachers and I sit quietly together beside pictures of Blanche and the altar. At twenty after two, we know that Blanche is gone.

Matt later tells me that Blanche fell into a deep sleep five minutes before the vet arrived, and so the transition from dreaming to dying was as natural as a leaf drifting down to the forest floor.

Matt says that after Blanche died Celeste immediately lay down next to him, her nose on his nose, her paw on his paw. She must have realized that it was her only chance to get close to Blanche without getting cuffed.

At my request, Matt wraps Blanche in my favorite cloud pajamas and puts him in the freezer so that we can say good-bye to him together before we take him to be cremated.

For four days after I return from teaching, I avoid opening the freezer; it is too spooky that Blanche is in there. But I also feel safe because he is still in the house. Then I begin getting worried that it isn't normal to keep my dead cat in the freezer and I should do something about it. I feel like Miss Havisham in *Great Expectations*, who was so broken-hearted when her fiancé jilted her that she lived the rest of her life in her wedding dress waiting for him to return.

One night, with Matt beside me, I lay Blanche on the floor in front of the fireplace. Although I can see that the life in his body is gone, he looks so much like himself — same whiskers, same paws, same bushy tail — that I can fool myself into believing he is sleeping. I try to say good-bye, but all I can do is rock and weep. The thought of never seeing him again, never being able to touch his fur, is unbearable. No matter what Matt says, no matter how much he holds me, comforts me, cries with me, I feel like a two-year-old who is being asked to give up the only love she ever knew. A dead body is better than no body. A frozen cat is better than no cat. I want to spend the rest of my life rocking on the floor, holding my dead cat in my arms.

Finally, after a week, I know I need help; I can no longer tell the difference between the part of me that believes there is nothing to live for and my self.

I take Blanche's body to Jeanne's office and together we unwrap the cloud pajamas. She isn't afraid of my sorrow; she doesn't think I am weird or sick, and within minutes, I stop being afraid as well. As I touch Blanche and feel the frozenness of his whiskers and eyelids and tail beneath my fingers, I begin to understand that he is really, really gone. This is only his body; whatever made him Blanche is not here anymore. There is nothing left to do but let this frozen body go.

Later that night, I meet Matt at the animal hospital, and we give the vet Blanche's body to be cremated. An elderly couple is sitting in the waiting room with their eleven-year-old sheltie named Poppy who has suddenly stopped eating. "It just happened yesterday," the woman says over and over. "She's such a good dog. She's our only child." I look at the woman's face; it is streaked with tears. Her white hair is rumpled and her purple velour sweater has two patches of dried egg yolk sticking to it. Her husband is sitting quietly beside her, holding her hand and shaking his head. I nod my head and cry with them.

Love is such a heartbreaker.

Eleven

In Pine Valley on *All My Children*, dead people spring back to life. Tad drowned and now he is back, Edmund showed up in the middle of his own funeral, and my mother and I read that Maria, who died in a plane crash three years ago, is going to reappear any day. In Pine Valley, dead doesn't mean gone; fatal gunshot wounds and funerals are merely mishaps, and no loss is ever permanent. These are some of the many reasons I watch *All My Children*.

Before my father's death, I kindasorta believed that people who died were like actors on soaps. They were milling in the wings of the theater, waiting to get enough standing ovations to walk back on and resume their parts. If I wanted them to be there badly enough, if I clapped and clapped, they'd show up. That is why, eight months after his funeral, I thought I saw my father.

I was back in Great Neck, walking across the street from the hotel in which I'd stayed while he was dying. From behind, I saw him in a wheelchair, hunched down, with his corn silk hair and his exposed, vulnerable neck. I started to walk very fast. I wanted to see his face one more time. I wanted him to look at me and say, "Hello, sweetheart. Howareya?" I wanted him to call me Wendell, Mabel, Mrs. Gabeanie. I wanted to hug him in the flesh, in his polo shirt with baggy corduroy pants and cherished Masonic ring on his left pinkie finger.

The nurse and the man turned the corner. I began to run after them, weeping, coat flying open. They turned into a hardware store, and I stood outside, hands cupped on the window, trying to see him. Finally, I heard an internal voice telling me to get a grip, my father was dead, this man was a stranger, he wouldn't recognize me, wouldn't call me Wendell. I removed my hands from the window, and as I did, his Masonic ring on *my* finger glinted in the light. It dawned on me — for the millionth time — that I would never see his face, hear his voice again, and I wanted to throw myself against a wall to stop the pain.

Now it is Blanche whom I am never going to see again. For the first few weeks after we bury him, I watch his video a few times a day. I cry when I wake up and he isn't here, cry when I eat breakfast and he isn't here, cry when I walk outside and he isn't under his favorite tree, and cry when I walk back inside and he isn't sleeping in his favorite rocking chair. I keep saying to Matt, "But how? How is it possible that he could be here and then not be?"

When I begin yelling at myself for sticking Blanche in the sunroom during his last few months of life, it reminds me of yelling at myself for not going to my cousin's wedding in Florida six months before my father was diagnosed with lymphoma, when I could have danced with him to "Strangers in the Night" one last time.

It is tempting to blame myself for Blanche's death, tempting to tell myself I could have done it differently. Sometimes the pain is so intense all I can do is make it worse by going over and over what I could have done, what I should have done, what a selfless person would have done. Blaming myself takes the focus off the pain; it gives me something to do besides feeling shattered.

I soon realize that I have two choices: allow the breaking or rail against it, allow what is already happening or create a drama on top of it. It occurs to me for the gazillionth time that Blanche or no Blanche, these are the only choices I ever have.

Twelve

When I was in eighth grade, my friend Dina Cooper lived above her family-owned funeral home. We played dress-up in cheap frilly skirts meant for someone else's dead aunt, passed the embalming room as we ran up the stairs. During games of hide-and-seek, we crouched behind open caskets with putty faces and powdered noses poking out. In those days, death seemed like a plush, velvety room you passed on the way to snacks of Keebler cookies and milk. Nothing about dying frightened me, except when it came to my father.

Before my father was ill, acknowledging death seemed like opening my heart to a stampede of elephants. But after he was diagnosed, when I leaned into the grief rather than denying it, when I allowed myself to be curious about the labyrinth of stories I'd been telling myself for half a century, when I practiced being fierce about staying in the present rather than wandering off into tragic drama, I began to feel as if I were gliding in a kind of vastness through which my father's life, and my life as I knew it, were being blazed. I began to relax. I stopped fighting, stopped resisting, stopped believing it would be better if he'd lived. It was as if I suddenly raised my head after a lifetime of hunching between my shoulder blades, and noticed things as they were, instead of being lost in the stories I'd constructed. I'd been so busy protecting my heart from being broken I didn't realize I was living barricaded, in a state of self-induced deprivation. I was like my grandmother, who wouldn't allow anyone to send her flowers, wouldn't let herself be thrilled by the lush, slow opening of a fist full of peonies, because she could not bear to see them die.

If anyone had asked me, even a year before his diagnosis, how my father's death would affect me, I would have said that I was going to be destroyed. But when he died, I felt as if my unspoken contract with him ended, and though the sadness of his passing affected me for months, I felt complete. My father's death was like a bonfire roaring through my life, burning so clean it left no traces of ash or smoke. When I think of him now I feel hugely grateful for what I *did* inherit from him — his passion for work, his willingness to belt out songs in the middle of anywhere, his unadorned love for Judy Garland. Although I still miss knowing I have a father, he was absent from my daily life for so many years that I don't miss him.

But Blanche is another story. By appearing

in cat form, he bypassed my beliefs, glided through my protective walls, and allowed me the sheer joy of loving without defense. A few weeks after his death, it occurred to me that I had always been able to love. I would have given up my life for my mother — surely that could be called love.

I had love the way water has wetness.

I can't help but have love; no one can.

Nevertheless, I will always miss the wheezy purr, the silkiness beneath my fingers, the be- mused look on his face. Even now, as I walk through the house, I often think I see the swish of his tail rounding a corner.

And, as it did with my father, awe accom- panies the grief. Blanche was here, and then he wasn't. Where did he go? Does anything ever die, or do we just stop being able to see its physical form?

The answer is: I don't know.

Before he died, I knew it all.

I knew I would be devastated by his death.

I knew my heart would close.

I knew I would be afraid to love anything that could die before me again.

But none of that is true. I am not devas- tated. I am very sad, and for a while I walked around in a regressed state, but the grief is moving through me like every other feeling has done. Eventually, I get up off the bed, wipe the hair off my wet, swollen face, and go into the kitchen to eat some popcorn.

Then I watch *All My Children*. I don't want to miss talking to my mother about Maria's return from the dead.

Nothing lasts, not even grief.

A recent newsmagazine featured an article about cloning pets. One man talked about certain animals that come along every now and then, special animals, once-in-a-lifetime animals — his dog was such an animal, and if he could clone her and have another one just like her, he said, it would be worth paying twenty thousand dollars. He never wanted to say good-bye to his dog.

I ask myself if, given the chance, I would have cloned Blanche.

As tempting as it is to think about having a replica of Blanche around, I would not have cloned him; the aftermath of surviving what I thought would destroy me, of having my heart broken ten thousand times, means there is nothing left to protect. I can have the whole world, every illuminated gorgeous fragile sunrise, every glimpse of a humming-bird, every morning I spend with Matt, every taste of food, every step, every breath. I can have the unrepeatable wonder of them, no holds barred — but only if I know I can lose them the next instant. It turns out that astonishing beauty co-exists with staggering pain. It's a feast here, it's always been a feast, and even dying is part of it.

In the end, both deaths — my father's and

Blanche's — did mean a kind of death for me. Because I could not be who I thought I was and live through their deaths, and because, in addition to the heaving grief, there were necklaces of whooshing, starry moments, I realize I am not who I thought I was and the world isn't the way I imagined it to be. I was living in a small corner of what is possible, constricted, constrained, locked in, caged by my beliefs about them, me, love, death. The heart of me is not a three-year-old, not even a forty-eight-year-old. If the worst possible thing that could happen does not destroy me, it means there is an ineffable, mysterious something that can never be destroyed.

Nothing is the way I thought it would be. It's better.

Thirteen

After Blanche's death, a friend sent me the Last Will and Testament that Silverdene Emblem O'Neill, Eugene O'Neill's dalmatian, wrote before he died. In it, Blemie (as he was known to friends and family) described proudly walking along the Place Vendôme in Paris, 1929, donning an Hermès collar; he also spoke of romping around chasing rabbits, and the good life he'd lived. At the end of his soliloquy, he beseeched his mistress and master not to grieve excessively for him, and to get another dog.

The very thought of being outclassed by a canine, especially in death, offended Blanche, and so he channeled the following words to me on the day before his memorial service. At his request, I now share these words with you, the reader.

You cannot see me splayed in the sunroom looking as if I am surfing on a wave of light, you cannot see me lapping up the dripping water in the bathtub, curled on the couch in the TV room, or snoring in the laundry basket. This deceives you into believing I am not here, but you are only looking with your physical eyes. Look again. Look with the eyes beneath your eyes, the quivering life beneath what you call your life. As you are beginning to discover, it is what you can see with those eyes that is most compelling. It's time to begin living the shimmery glimmery sunlit life you gave me, but haven't let yourself fully inhabit.

Everyone knows I had a better life — and death — than most people on the planet. Between the acupuncturists and the psychics, being hand-fed and carried everywhere, having mice heads to eat, dogs to chase, fences to jump, and corn on the cob to nibble on, there was nothing the physical world didn't offer for my pleasure. And who wouldn't want a death like mine — carried around in a cashmere snuggly, touched sweetly until my last breath, with a Zen priest and a pearly godmother chanting softly beside me.

All that was good, but the pleasures of the physical world — jeweled collars and sparkly necklaces, white downy blankets and dried salmon flakes — were not the real treasure. It was the love, it was always the love. It was the fact that you delighted every time you saw me. Every time for seventeen and a half years, I knew that just by walking into a room, your heart would fling out streamers of joy, so I kept walking so that your heart could keep flinging, and I kept putting my paws on your face so that your body could keep relaxing, and I kept purring so you would know there was safety in this world, but it wasn't me any more than it was the jeweled collars.

It was you.

It was always you.

You used to mistake the symbol of the treasure for the treasure, the marker for the thing itself, the gift from God for God, the statue of Buddha for his infinite vastness. As if all you could possibly hope for was a thing you could touch, a token, rather than all of shining existence. Since you hadn't let yourself know that shimmering fully, you kept turning to what reminded you of it — glitter and baubles and sparkles. As if having those was having the real thing. As if that was the best you could do.

It was time for me to go. I told you I would stay until you were strong enough to live without me, and I did, and you are. Until your heart spread like dragonfly wings, until you didn't need me to know you had a heart. As long as I was in a physical body, you relied on me. You believed I was the locus of that love. Now you can find out for yourself what is true.

Do not grieve for me. I am in a place where tuna fish juice flows like water, where I can jump like the wind and every place is silky and sunny. If you must, grieve for what you won't allow yourself to have. Grieve for all the ways you separate yourself from this radiance, from lying down in a patch of sun at two o'clock on any old day, from knowing you are beloved on the earth.

Acknowledgments

I don't know where to begin. I had so much help during the writing of these pages — and the manuscript took so many unexpected turns — that without the following people, this book would not be this book:

Peter Guzzardi cared about every word right along with me, and remained steadfast in his belief that seemingly unrelated themes could be woven together in a cohesive whole. Given the solitary nature of writing, it is his great gift that I felt partnered by him.

Mary Jane Ryan's relentlessly positive encouragement and wise book sense got me back on my feet again.

Shaye Areheart, my publisher at Harmony, is a powerhouse of energy, understanding, and laser-like intelligence. Together with the fabulous Kim Meisner — who sat and labored with me page by page — their insights revealed an unexplored dimension, both in my writing and in my life.

Angela Miller, my trusted friend and crackerjack agent of twenty-two years, always manages to get each book into the right publishing hands.

Taj Inayat roamed the skies with me, often

understanding what I wanted to say before I did.

Annie Lamott's kind and brilliant reading opened the way for radical revision.

Kim Rosen's generosity with her time, friendship, and gorgeous words were a continual source of joy.

Sandra Maitri's precision about spiritual and psychodynamic issues allowed me to deepen my own understanding.

Jace Schinderman, Chohan Neale, Francie White, Blanchefleur Macher, and Howard Roth helped refine the manuscript with their perceptive readings.

Anna Schmitz brainstormed my favorite title with me, and her husband, Jean-Claude West, knows everything about (almost) everything, for which I am continually thankful.

I can't imagine my life without Maureen Nemeth, who, with her astute observations, loyalty, and humor, keeps the rest of my life intact when I disappear into writing for weeks at a time.

On Blanche's behalf, I want to thank: Sally Blumenthal McGannon for ignoring me when I said I was unfit to be a cat mother; Dr. Cheryl Schwartz for the amazing, respectful, tender way she engaged with Blanche, and for teaching me to do the same; Paul Rosenblum for recognizing Blanche as the kingly character he was; Karen Johnson for staying with me during that last, heart-breaking phone call; Catherine Bacon for blazing a path of lush beauty and unrestrained four-legged love; and Stan Mellin for his thoughtfulness in sending the O'Neill eulogy that prompted Blanche to speak for himself.

I am blessed to have Glenn Francis, the world's most perfect meditation teacher, as a spiritual friend and colleague. Our co-teachers, Premsiri Lewen, Sara Hurley, Gitte Dobrer, and Menno de Lange each contributed their fields of mastery to our work together, and I have benefited greatly from their wisdom and clear thinking.

My readers and workshop participants continually inspire me with their courage to lean into their fears. And it is my great privilege to work with my retreat students as they allow themselves to discover what never dies.

To Jeanne Hay, for her luminous face of Love, I am forever, unspeakably grateful. Because of her, not only does this book exist, but I exist as my Self as well.

Catherine Ingram ushered me into a palace of sky light.

Hameed Ali stands as a human reminder to the groundless ground that is the source of it all.

Finally, and immensely, for reading each word a hundred or more times, and for continuing to beam his living light on me day after day, I thank Matt Weinstein. Blanche opened the door to my heart, and you walked through it.

About the Author

Geneen Roth is the author of seven books, including the *New York Times* bestseller *When Food Is Love*. She has appeared on *The Oprah Winfrey Show*, *Good Morning America*, *20/20*, and many other national television shows, and her work has been featured in numerous publications. She lives in northern California and maintains an active lecture and workshop schedule. Visit her website at www.geneenroth.com.